Climate Adaptation Finance and Investment in California

This book serves as a guide for local governments and private enterprises as they navigate the unchartered waters of investing in climate change adaptation and resilience. This book serves not only as a resource guide for identifying potential funding sources but also as a roadmap for asset management and public finance processes. It highlights practical synergies between funding mechanisms, as well as the conflicts that may arise between varying interests and strategies. While the main focus of this work is on the State of California, this book offers broader insights for how states, local governments and private enterprises can take those critical first steps in investing in society's collective adaptation to climate change.

Jesse M. Keenan is a faculty member at the Graduate School of Design at Harvard University where he researches and teaches courses in climate change adaptation and the built environment. Keenan concurrently serves as a Research Advisor for Climate Adaptation Finance to the Governor's Office of Planning and Research for the State of California and as a Visiting Scholar at the Federal Reserve Bank of San Francisco.

Routledge Focus on Environment and Sustainability

For more information about this series, please visit: www.routledge.com/
Routledge-Focus-on-Environment-and-Sustainability/book-series/RFES

Climate Adaptation Finance and Investment in California

Jesse M. Keenan

LONDON AND NEW YORK

First published 2019
by Routledge
2 Park Square, Milton Park, Abingdon, Oxon OX14 4RN

and by Routledge
52 Vanderbilt Avenue, New York, NY 10017

Routledge is an imprint of the Taylor & Francis Group, an informa business

British Library Cataloguing-in-Publication Data
A catalogue record for this book is available from the British Library

Library of Congress Cataloging-in-Publication Data
A catalog record for this book has been requested

ISBN: 978-0-367-02607-3 (hbk)
ISBN: 978-0-429-39875-9 (ebk)

Typeset in Times New Roman
by Apex CoVantage, LLC

State of California
Edmund G. Brown Jr., Governor

Governor's Office of Planning and Research
Ken Alex, Director

Contents

Acknowledgments

I would like to acknowledge the tireless efforts of Nuin-Tara Key and the team at the Governor's Office of Planning and Research for their support in researching and producing this book. I would also like to thank my research team at Harvard University, including Susanna Pho (Lead Visual Designer), Carolyn Angius (Research Supervisor), Pamela Cabrera (Illustrator), Anurag Gumber (Researcher) and Jenny Fan (Graduate Researcher). I would also like to acknowledge Louise Bedsworth, Ken Alex, Brian Strong, John Moon, Laura Choi, Naomi Cytron, Lizzy Mattiuzzi, Laurel Gourd and the Federal Reserve Bank of San Francisco for their support in researching this book. I am also grateful to Governor Jerry Brown, Sam Higuchi, Ann Kosmal, Michael Bell, Bill Solecki, Cynthia Rosenzweig, Peter Huybers, Dan Schrag, Jim Clem, Radley Horton, Rachel Minnery, Judge Alice Hill, Joseph P. Kennedy II, Brian O'Connor, Peter Smith, Chris Flavelle, David Martin, Doug Parsons, Steven C. Cronig, Bud and Leslie Bell, David F. Jones, Eddie Lopez, Mike and Sue Keenan, Nathaniel and Lindsey Keenan and Kristen Keenan for their ongoing inspiration in advancing a world that is more fair, just and knowledgeable in the face of climate change. This book was supported by generous grants from the Harvard University Climate Change Solutions Fund and the Federal Reserve Bank of San Francisco.

Preface

This book serves as a guide for local governments and private enterprises as they navigate the unchartered waters of investing in climate change adaptation and resilience. This guide is intended to serve not only as a resource guide for identifying potential funding sources but also as a roadmap for asset management and public finance processes. It does so by framing what questions should be asked today of what might happen tomorrow. It highlights practical synergies between funding mechanisms, as well as the conflicts that may arise between varying interests and strategies. In particular, this guide provides available methods for ensuring that social equity considerations help shape fair and equitable investments.

While this guide is intended to address an audience with a working knowledge of finance and investment, it also offers an accessible perspective on basic concepts and ideas that challenge the assumptions of conventional investing. To this end, asset managers, risk managers, underwriters, project managers, planners, civic advocates and elected officials will all find value in this guide, as they seek to invest in projects and programs that address climate change impacts. This guide is part of California's commitment to providing the most up-to-date knowledge for guiding the state's climate future. Together, the public, private and civic sectors can work together to maintain an intelligence about emerging markets, instruments and models that serves to not only manage risks but also to take advantage of the many opportunities for civic investment.

Readers may use and interpret the information contained in this guide at their sole discretion. Nothing in this guide should be construed as establishing any legal mandates or required actions.

Acronyms and abbreviations

AAL	Average Annual Losses
AB	California Assembly Bill
ACEP	Agricultural Conservation Easement Program
AHSC	Affordable Housing and Sustainable Communities Program
ASCE	American Society of Civil Engineers
AWWA	American Water Works Association
BFC	Blue Forest Conservation
BID	Business Improvement District
BUILD	Better Utilizing Investments to Leverage Development Grants
CAB	Capital Appreciation Bond
CAGHAD	California Association of Geological Hazard Abatement Districts
CALED	California Association of Local Economic Development
Cal EPA	California Environmental Protection Agency
CAL FIRE	California Department of Forestry and Fire Protection
Cal OES	California Office of Emergency Services
Caltrans	California Department of Transportation
CAP	Continuing Authorities Program
CARB	California Air Resources Board
Cat Bond	Catastrophe Bond
CBA	Cost-Benefit Analysis
CBI	Climate Bonds Initiative
CCC	California Coastal Conservancy
CCFRF	Community Choice Flood Risk Financing
CCI	California Climate Investments
CDBG	Community Development Block Grant
CDC	Centers for Disease Control and Prevention
CDFA	California Department of Food and Agriculture
CDFI	Community Development Finance Institution
CDFW	California Department of Fish and Wildlife

CEA	Cost-Effectiveness Analysis
CEC	California Energy Commission
CEQA	California Environmental Quality Act
CFD	Community Facilities District
CFIP	California Forest Improvement Program
CLEEN	California Lending for Energy and Environmental Needs Center
CMAQ	Congestion Mitigation and Air Quality Improvement
CNRA	California Natural Resources Agency
CPUC	California Public Utilities Commission
CTA	Chicago Transit Authority
CTC	California Transportation Commission
DHS	United States Department of Homeland Security
DOE	United States Department of Energy
DPR	Department of Parks and Recreation
DWR	Department of Water Resources
EEM	Environmental Enhancement and Mitigation
EIFDs	Enhanced Infrastructure Finance Districts
EPA	Environmental Protection Agency
EPIC	Electric Program Investment Charge
ESCO	Energy Service Companies
EV	Expected Value
FEMA	Federal Emergency Management Agency
FMA	Flood Mitigation Assistance Program
FSB	Financial Stability Board
FTA	Federal Transit Administration
GARI	Global Adaptation & Resilience Investment Working Group
GGRF	Greenhouse Gas Reduction Fund
GHAD	Geological Hazard Abatement District
GIIN	Global Impact Investors Network
GIIRS	Global Impact Investment Rating System
GO	General Obligation
GSA	Government of South Australia
HCD	California Department of Housing and Community Development
HMG	Hazard Mitigation Grants
IBank	California Infrastructure and Economic Development Bank
ICARP	California Integrated Climate Adaptation and Resiliency Program
ICMA	International Capital Market Association
IFD	Infrastructure Finance District
IPCC	Intergovernmental Panel on Climate Change

IRIS	Impact Reporting and Investment Standards
IRWMP	Integrated Regional Water Management Plan
ISRF	Infrastructure State Revolving Fund
LAC	Los Angeles County Regional Park and Open Space District
LAO	Legislative Analyst's Office
LCCA	Lifecycle Cost Accounting
LCTOP	Low Carbon Transit Operations Program
LEP	Limited English Proficiency
LGBTQQ	Lesbian, Gay, Bisexual, Transgender, Queer and Questioning
LHMP	Local Hazard Mitigation Planning
LOCA	Localized Constructed Analogs
MCA	Multi-Criteria Analysis
MSRB	Municipal Securities Rulemaking Board
MTC	Metropolitan Transportation Commission
NAIC	National Association of Insurance Commissioners
NDRC	National Disaster Resilience Competition
NEPA	National Environmental Protection Act
NFIP	National Flood Insurance Program
NIBS	National Institute of Building Sciences
NIST	United States National Institute for Standards and Technology
NOAA	National Oceanographic and Atmospheric Agency
NPV	Net Present Value
O&M	Operations and Maintenance
OES	California Governor's Office of Emergency Services
OGALS	California State Parks' Office of Grants and Local Services Program
OM&R	Operation, Maintenance and Repair
OMB	Office of Management and Budget, President of the United States
OPC	California Ocean Protection Council
OPR	Office of Planning and Research, Governor of the State of California
PA	Portfolio Analysis
PACE	Property Assessed Clean Energy
PDM	Pre-Disaster Mitigation Program
PFA	Public Financing Authority
PG&E	Pacific Gas and Electric
PV	Present Value
RBD	*Resilient by Design: Bay Area Challenge Finance Advisory Team*
RDM	Robust Decision-Making
ROA	Real Options Analysis
ROI	Return-on-Investment

RRAP	Regional Resilience Assessment Program
SAD	Special Assessment District
SALC	Sustainable Agricultural Lands Conservation
SB	California Senate Bill
SEC	Securities and Exchange Commission
SGC	Strategic Growth Council
SHMP	State Hazard Mitigation Plan
SLR	Sea Level Rise
SPARCC	Strong, Prosperous and Resilient Communities Challenge
SWCCP	South West Climate Change Portal
SWF	Social Welfare Functions
SWRCB	California State Water Resources Control Board
TAC	Technical Advisory Council
TCC	Transformative Climate Communities
TIF	Tax Increment Financing
TIRCP	Transit and Intercity Rail Capital Program
TRB	Transportation Research Board
U(∂Y)	Incremental Utility
UCSD	University of California San Diego
USACE	United States Army Corps of Engineers
USBR	United States Bureau of Reclamation
USDA	United States Department of Agriculture
USDOT	United States Department of Transportation
USFS	United States Forest Service
USFWS	United States Fish and Wildlife Service
USGCRP	United States Global Change Research Program
USGSA	United States General Services Administration
USHUD	United States Department of Housing and Urban Development
VaR	Value-at-Risk
WCB	California Wildlife Conservation Board
WRI	World Resources Institute

Figures, tables and boxes

Figures

Tables

Boxes

1   Introduction

This book serves as a guide for local governments and private enterprises as they navigate the unchartered waters of investing in climate change adaptation and resilience. Local governments and private enterprises in the State of California have made tremendous strides in developing an adaptive capacity for addressing the current and future impacts of climate change. Both the public and private sectors have been united in their challenge to not only conceptualize the economic consequences of climate change but also to develop a practical set of methodologies and criteria for evaluating investments undertaken in the name of adaptation and resilience. Through successive adaptation plans and updates for nearly the past decade, the State of California has made advances in framing investment challenges and interventions in everything from transportation finance to disaster recovery grants and from life-cycle asset management to structured finance (CNRA 2009, 2016, 2018a). The 2018 update to the state adaptation plan has called for not only advancing innovation in financing models but also the incorporation of climate adaptation into existing funding sources (CNRA 2018a, p. 88).

Advanced in coordination with the California Integrated Climate Adaptation and Resiliency Program (ICARP) Technical Advisory Council (TAC) and the Federal Reserve Bank of San Francisco, this guide seeks to provide insight into how local governments, as well as private enterprises, may strategically develop financial models based on a variety of funding sources, analytical methods and strategic motivations. In this regard, the challenge is not only to fund "climate" projects but also to fund every day projects that seek to incorporate some aspects of resilience and/or adaptation performance into their design standards and investment underwriting criteria. From another perspective, this guide helps provide a methodology for underwriting resilience and adaptation considerations in projects that serve a variety of interests and social equities over a variety of time horizons. Consistent with the principles of the ICARP-TAC and its authorizing legislation, it is incumbent upon stakeholders to evaluate the distributive costs and benefits that shape the social, economic and environmental welfare of vulnerable communities. As such, this guide provides methods for ensuring that social equity considerations help shape fair and equitable investments.

The broad intent is to develop a sensitivity in underwriting and managing investments that provide transparency for investors and the general public about the nature of trade-offs by and between different options and strategies. As will be discussed, these trade-offs may be between investing in short-term resilience and long-term adaptation or simply between the conflicts that may arise between the built and natural environments. The intent is to empower local governments to not only develop innovative finance models but also to communicate the value of such investments to the general public, as well as those who are the stewards of managing assets. As such, this guide attempts to think beyond the immediacy of return-on-investment (ROI) analysis in favor of a multitude of quantitative returns and qualitative benefits.

How to use this guide

This guide is intended to provide a survey of issues, considerations and sources of funding that can help guide strategies and tactics for investing in adaptation and resilience in California. While this guide is primarily oriented for asset management, public accounting, risk management and transactional finance actors within local governments, it may also be insightful for actors engaged in community development and investment; state agencies interested in developing adaptation finance products or conduits; and, private sector financiers and underwriters who recognize the opportunities associated with responsible investing in climate change. The intent of this guide is not to provide a prescriptive pathway or underwriting process but to challenge the assumptions and values of existing modes of analysis, as well as to highlight novel ideas and developments that are likely to have bearing on future investments. By orienting adaptation finance to existing conventions, the guide is intended to reflect on the proposition that the "finance" in adaptation finance is relatively straightforward. The more fundamental challenge is in identifying the sources of funding that will allow for new capital stacks that account for the divergent interests and returns associated with a new form of investment.

Momentum shaping adaptation finance

This guide builds off of the work of a variety of California-specific resources that have sought to mobilize greater analytical sensitivity and issue awareness associated with current and future climate adaptation investment challenges. Pursuant to Executive Order B-30–15, the *Planning and Investing for a Resilient California: A Guidebook for State Agencies* (OPR 2017) provides a generalizable process for assessing climate impacts and risks in projects, as well as a variety of conceptual models for advancing data-informed decision-making processes. This accessible process is contextualized within permitting and economic analysis tasks within existing policy regimes, including the California Environmental Quality Act (CEQA) and National Environmental Protection Act (NEPA). Most importantly, this guide provides specific benchmarks, climate data sets and methodologies for accessing and utilizing climate services, including downscaled models that are most appropriate for California. This is an important resource because it sets the standards for normalizing data that allows for an assessment of risk and opportunity by and between similarly situated alternative investment options. In addition, this also serves to create a benchmark for best practices consistent with local planning and design standards that must independently have concurrency with investment underwriting criteria.

Adaptation finance has also been the object of research through the Fourth California Climate Assessment. As will be discussed in more detail in Chapter 2, *Adaptation Finance Challenges: Characteristic Patterns Facing California Local Governments and Ways to Overcome Them* (Moser, Ekstrom, Kim, & Heitsch 2018) has provided significant insight into not only understanding potential innovations but also how those innovations may be constrained by virtue of institutional, administrative, communication and legal barriers. More fundamentally, however, the research highlighted that "there [are no] estimates available for California (or any state) for how much money has been spent on adaptation to date and how much more is needed to support local adaptation" (id., p. 2).

With this uncertainty in mind, it can be argued that, without even knowing the range for the overall projected need, it is necessary to shift the framing from an optimization (e.g., economic growth vs. averted losses/adaptation costs) for the allocation of limited resources to a qualitative framing that seeks to develop a robust capacity for mainstreaming within a variety of sector-specific capacities for funding adaptation interventions (Dittrich, Wreford, & Moran 2016; Heal 2017). While models do exist for estimating a range of economic losses from climate change (Hsiang et al. 2017; Tol 2018), this is not the same as estimating project life-cycle impacts that may highlight asset impairment that otherwise suggests increased or novel risks and opportunities associated with collateral valuation and credit risk (FSB 2017a; Ernst & Young 2017).

Pursuant to Executive Order B-30–15, state agencies are already required to evaluate full life-cycle cost accounting (LCCA). Further, pursuant to state guidance, this LCCA should be inclusive of not only design and performance criteria that reflect a changing climate but also an incorporation of costs associated with the operations, maintenance and repairs (OM&R) of those assets (OPR 2017). This should include estimates for performance under conditions of climate stress and resilience costs associated with continuity and recovery following extreme events or shocks. In addition to these policy considerations, model uncertainty also raises the question as to whether deterministic net present value (NPV) methodologies such as cost-benefit analysis (CBA) are adequate in light of the proposition that many critical conditions and assumptions are likely to change following an initial underwriting (Watkiss, Hunt, Blyth, & Dyszynski 2015). For instance, how would the NPV payback period for a resilience investment change within a LCCA, if temperatures increase faster than was modeled?

While this guide will primarily build upon conventional financial methodologies, it seeks to address these model uncertainties by qualitatively challenging some of the assumptions that set-up quantitative sensitivity and scenario analyses necessary to account for uncertainty, deep uncertainty and a general lack of knowledge or awareness (Hallegatte, Shah, Brown, Lempert, & Gill 2012). Aside from model uncertainties, the more fundamental uncertainty relates to the timing, frequency, magnitude and severity of direct and indirect climate change impacts on assets, operations and programs. To account for this range of uncertainty, real options approaches are increasingly being utilized to account for the value of flexibility or delaying decisions in the face of such uncertainty (Buurman & Babovic 2016). However, for many local governments with less capacity for sophisticated financial analysis, there is a necessity to keep the analysis simple and challenge the assumptions. These assumptions include everything from operations and maintenance liabilities to rates of material and performance degradation under extreme environmental stress. Whether it is a matter of performance or financial liability, nearly every assumption relating to the financing and operations of assets may be challenged by direct or indirect climate change impacts.

What is resilience and adaptation?

The concepts of resilience and adaptation have a variety of meanings and applications across a variety of sectors, practices and stakeholders. Pursuant to the National Climate Assessment, the U.S. government defines resilience as "[a] capability to anticipate, prepare for, respond to, and recover from significant multi-hazard threats with minimum damage to social well-being, the economy, and the environment" (USGCRP 2018a). Adaptation is defined as "an adjustment in natural or human systems in response to actual or expected climatic stimuli or their effects, which moderates harm or exploits beneficial opportunities" (OPR 2017, p. 6). However, these definitions are often viewed as being somewhat abstract for purposes of guiding planning and decision-making (Larkin et al. 2015). OPR defines climate resilience as "the capacity of any entity – an individual, a community, an organization, or a natural system – to prepare for disruptions, to recover from shocks and stresses, and to adapt and grow from a disruptive experience" (OPR 2017, p. 61). In this regard, resilience is understood to advance a smarter capacity to prepare and recover based on continual learning, flexibility and adjustment.

But, why are these concepts of resilience and adaptation important to finance and investment? They are important because they represent processes that help frame options and trade-offs that are central to an investment analysis. In order to apply these concepts, it is necessary to distill them to their most basic meanings. There are multiple categorical variants of resilience, including ecological, socioecological, urban, disaster, engineering and community resilience (Meerow, Newell, & Stults 2016). For the purposes of this guide, it is engineering resilience and community resilience that have the most bearing for an investment analysis (Davidson et al. 2016). Technical definitions aside, engineering and community resilience are essentially about a capacity of an engineered system or a community of people to absorb and recover from shocks and to learn from that process of recovery so that future shocks aren't so disruptive or catastrophic.

However, for purposes of investment, keeping it simple is paramount. As such, resilience is fundamentally about preserving the relative status quo through forms of recovery (Keenan, King, & Willis 2015). Organizations, systems and assets may "build back better" but the status quo is maintained by the mere fact that one is building back at all. In practice, this is primarily understood in terms of short- to mid-term time horizons in response to shocks and sometimes stresses. Although technically distinct, this is very often referenced in terms of infrastructure protection and socioeconomic stabilization. By contrast, risk mitigation is about preventing the risk from impacting a system or community. Figure 1.1 provides a graphic representation of the conceptual relationship between resilience, risk mitigation and adaptation. Given the same perturbation (e.g., extreme heat, sea level rise), hazard mitigation prevents the system or community from being impacted, while resilience allows the same group to maintain performance despite being directly impacted.

In this regard, engineering resilience is measured by the time and cost associated with the capacity to recover to a pre-shock state. For purposes of asset management and investment, it is important to note that aging effects of engineered systems (e.g., gray infrastructure or buildings) will alter a post-event state. In this regard, resilience may operate to return an asset to

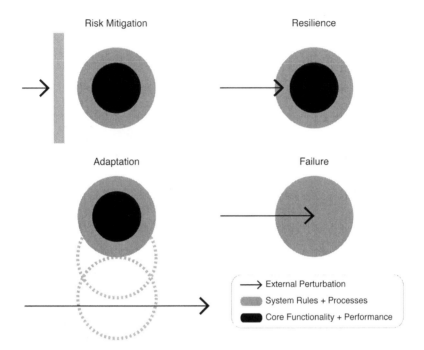

Figure 1.1 Conceptual relationship between risk mitigation, resilience, adaptation and failure

a state that includes some measure of accelerated degradation and/or depreciation within its normal life-cycle by virtue of the fact that an asset that has endured some degree of stress that may, in part, be attributable to climate change impacts (Kurth, Keenan, Sasani, & Linkov 2018). As represented in Figure 1.2, the resilience of a system has a threshold and beyond that threshold one either fails or adapts. As highlighted in the figure, some resilience investments may be maladaptive given the relative disparity caused by ongoing climate stressed-induced aging. In some cases, replacing an asset or a component of an asset may be a superior proposition to investing in its resilience. In this case, the analysis must center on the direct and indirect costs associated with non-performance during which time the asset or component fails and is replaced.

By contrast, the most simplified meaning of adaptation is one that speaks to the capacity of an engineered system, organization, community or individual to transform to an alternative domain of operation (Adger, Arnell, &

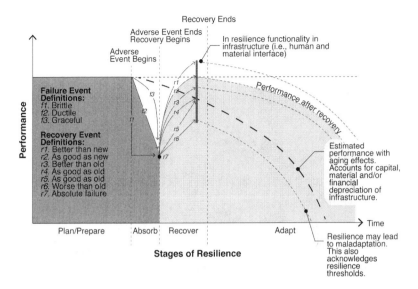

Figure 1.2 Engineering resilience curve relationship to adaptation in built assets

Source: Adapted from Kurth et al. (2018).

Thompkins 2005). The key concepts are "capacity" and "transformation" – doing things differently to achieve a similar function or purpose which "moderates harm [and] exploits . . . opportunities" (OPR 2017, p. 6). If resilience is about managing risks to preserve the status quo, then adaptation is about a capacity to do things differently to take advantage of opportunities in light of the fact that preserving the status quo is no longer sustainable. As highlighted in Figure 1.1, the adaptation of the system or community means that its transformation has put it out of harm's way – at least to this one perturbation.

As represented in Figure 1.3, adaptation is often measured as the optimal balance between costs and the benefits of avoided impacts and associated losses. What is not represented in this figure is the opportunity to capture the up-side of adaptation investments when capital is otherwise reorganized. By extension, it has been observed that Wall Street is already making money from climate change adaptation through the modification of market niches, supply chain optimization, informational asymmetry and the indirect benefits from organizational capacity building and intelligence gathering (Keenan 2015a; White & Grantham 2017).

For purposes of strategic investment, the distinction between resilience and adaptation is important because it highlights the tension between

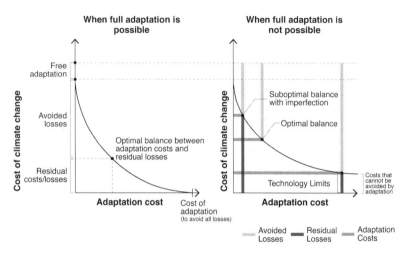

Figure 1.3 Balancing adaptation costs and benefits/avoided losses
Source: Adapted from IPCC (2014).

investment and disinvestment; short-term and long-term biasing; and the inherent trade-offs for the realization and recognition of value accruing at different times to different parties. For instance, there are many examples where resilience doesn't make a lot of financial sense because the costs are too high relative to what one is trying to preserve. Hence, a resilience investment can be internally maladaptive. For instance, resilience investments that seek to preserve elements of a system that are at the end of their useful life are likely maladaptive. The classic example revolves around the costs of elevating an older home that lacks any intrinsic historic value and is otherwise near the end of its useful life. All things being equal in terms of social costs, if the house is valued at $100,000 and it costs $150,000 to elevate the house, then it is not likely to be a sound investment.

Both concepts are plagued by the subjectivity of what objects or people are benefited by resilience or adaptation investments and who bears the burden of the costs and unintended consequences. Therefore, any analysis should be as specific as possible when applying these concepts. That specificity should reflect the exact nature of the risk or hazard; the physical or geographic limitations; and, the associated life-cycle constraints and time-horizon parameters for anticipated benefits. Thereafter, one has to ask whether resilience under these limited conditions is either internally maladaptive to the economics of the investment or whether it is maladaptive in terms of conflicting with other public policies or values. As such, both of these concepts have the potential for both conflict and synergy.

In practice, there is often a disconnect between who pays and who benefits. It is entirely dependent on the time horizon and the nature of the objects or beneficiaries of resilience and/or adaptation. Resilience to one population may be maladaptive to another, as is the case when municipal taxpayers are asked to bear the costs of protecting a small number of luxury coastal properties. Likewise, adaptation to one population may undermine the resilience of another population, as is the case when community resilience for a vulnerable population is weakened when strategic obsolescence in infrastructure is institutionalized in disadvantaged communities otherwise challenged by environmental justice. Understanding these conceptual and practical conflicts is central to not only framing trade-offs but also understanding the nature of social equity in shaping adaptation investments.

Box 1.1 Public investment in a living shoreline

By example, it can be assumed that there are at least two investment options for addressing coastal flooding in a neighborhood of low-income property owners. It has been determined that the costs of inaction are too great. The engineering resilience option would be to build a sea wall that can withstand incremental sea level rise (SLR) up to a certain threshold. The transformative adaptation option may be to gradually move the vulnerable populations and de-invest in existing sea wall maintenance and capital improvements. Assuming for the purposes of this example that the externalities for both options are fully internalized, there are two possible outcomes. In this case, the time horizon is defined by the term of a general obligation (GO) bond and the discount rate is the bond rate. The first option suggests that the long-term costs for maintaining the sea wall, together with a reduction in risk exposure, represent a higher positive NPV than the alternative costs of relocation and losing a tax base. The second option is that, given the relatively limited impact of the property on the tax base and the high cost of building a new sea wall, it makes more sense to invest in relocating households to higher ground through a combination of buyouts and the development of affordable housing on publicly purchased land within the existing jurisdiction. *See* Figure 1.4.

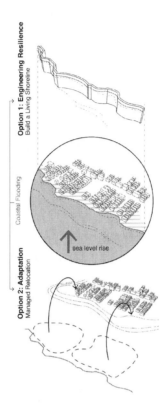

Figure 1.4 Public investment in living shoreline

Box 1.2 Private investment in a hospital

By example, it can be assumed that there are at least two options for assessing whether to hold or sell a small privately owned and aging hospital located on a fire-prone ridge. With climate change, forest fires are anticipated to be more frequent and investors must develop an investment strategy for managing a local hospital that recently had to temporarily shut down operations due to the proximity of a forest fire. The resilience option would be to invest in a business continuity-focused strategy that allows the hospital to maintain critical operations without the necessity to evacuate patients, although this strategy would not address the hospital's capacity to deliver high margin

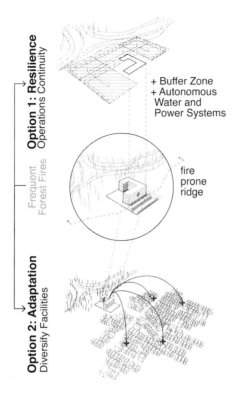

Figure 1.5 Private investment in a hospital

outpatient care services. The engineering resilience interventions include buying land around the hospital to provide a buffer zone and installing autonomous power and water facilities. The transformative adaptation strategy is to de-invest in the facilities and convert the hospital into a more profitable outpatient service facility. For the resilience strategy, the price of the land and the cost of the autonomous facilities may or may not yield an adequate ROI through risk reduction (e.g., lower insurance premiums, self-insured exposure) consistent with a payback period aligned with the probability occurrence of future events in terms of the average annual losses (AAL) from fire events. In one scenario, a low-probability and low-impact event may not be regular enough to justify such capital investments in resilience. In a high-probability and high-impact scenario, the owners may wish to extract as much cash-flow out of the facility as possible through outpatient services and to reduce their overall investment footprint through the liquidation of equipment and the reduction of their labor force. This may be an adaptive economic decision in favor of the economic viability of the hospital as a portfolio of assets, but it may be maladaptive for the communities that it serves in that the resilience of such communities may be undermined by a lack of access to inpatient services. *See* Figure 1.5.

From a practical point of view, underwriters and analysts have to ask themselves: is this resilience investment that helps population X in the short term going to be maladaptive to population X, Y or Z in the long term? In the world of finance, the populations largely revolve around borrowers, investors and lenders. As such, credit risks may arise over different time horizons and may originate among a variety of different classes of taxpayers, borrowers and guarantors. Public borrowers create path dependencies for every investment that they make that will dictate their capacity to make future investments in climate change resilience and/or investments. For instance, underappreciating life-cycle OM&R costs for large resilience investments may significantly impair the financial capacity and flexibility to make future investments. There is a risk that large short-term resilience investments will limit the adaptive capacity of borrowers to accommodate unknown future stresses and shocks, such as those impacts that may arise from rapid ice melt scenarios. Conversely, inadequate resilience investments may impair assets and tax-bases in a manner that limits their borrowing capacity for adaptation in the future. In these cases, expected value (EV) functions may be appropriate for estimating liability and probability occurrence of events within a broader enterprise risk management strategy.

However, if there is a tendency to optimize investment allocations in favor of the resilience of economic productivity and tax-bases, then the implications for disadvantaged communities and vulnerable populations may be dire. This friction will serve as a central point of contention for the timing and allocation of public investments in the future. With both resilience and adaptation, there will be economic winners and losers. The challenge is to balance short- and long-term interests and to maximize the range of beneficiates over the greatest amount of time. From the point of view of private sector actors, this may appear to be a set of considerations that are external to the transactions associated with project finance. To the contrary, credit rating agencies are now keenly aware of these trade-offs and are prepared to account for such investments and path dependencies in accounting for municipal credit ratings (Moody's Investor Services 2017; S&P Global 2017, 2018). Unfortunately, credit rating agencies may prefer that local governments invest in resilience investments that preserve a tax base – with little regard for social equity considerations.

For both resilience and adaptation, it is key to remember that each concept represents a process and not necessarily an outcome. While there may be intermediate outcomes that represent absolute adaptation or resilience interventions, the dynamic nature of climate requires an ongoing investment in the intelligence and resources necessary to promote an adaptive capacity for the knowns and unknowns of climate change. From an institutional perspective, this is the essence of adaptive management. Because of the dynamic nature of climate change, rules and assumptions are rarely stable for long. For purposes of investment, these concepts should be conceived of as emergent strategies that require regular ongoing intelligence and monitoring. As referenced in *Planning and Investing for a Resilient California: A Guidebook for State Agencies*, there are several key steps for incorporating climate change into a general analysis (OPR 2017). The first step is to identify how climate change could or may affect a project. This includes identifying possible direct and indirect impacts in terms of the scale, scope and context of an investment. In a best-case scenario, there are probabilities for the occurrence and intensity of certain impacts to help guide the costs of action and inaction. In other cases, there may be significant uncertainty concerning the time and/or intensity of an impact, as well as the consequences of direct and cascading hazards that may be associated with such impacts.

In both cases, scenario analysis and stochastic modeling may be useful in understanding a range of potential investment options, including the option of inaction or de-investment. Stochastic modeling is an analytical technique that allows for estimating probabilities based on random variation of one or more variables or inputs over a period of time. It is used to simulate outcomes of various scenarios that might occur in the future. Such a technique may demonstrate a high-probability, high-impact event that may support a decision that de-investment may be an appropriate adaptation option. In either event, it is the identification of a range that is most important for informing decision-making. Attempts at a binary optimization centered around a singular number are likely based on assumptions that are more often than not the qualitative purview of the judgment of investment professionals and elected officials. Therefore, it is critical to develop a set of criteria and values that represent the risk-tolerance, public policy ambitions and investment return benchmarks that collectively serve as boundaries to a range of possible financial options. Chapter 2 will provide additional guidance on how best to approach asset management considerations, and

Chapter 4 will provide guidance on how to evaluate different funding and financing options.

Challenges and opportunities of adaptation finance

Adaptation finance is a rapidly emerging area of interest as a result of increasing awareness of the range of ongoing and future climate change impacts and the costs associated with mitigating and adapting to those impacts. As previously referenced, Moser et al. (2018) provide an exhaustive review of various challenges facing local governments in their capacity to assess, access, utilize and manage funding for purposes of making investments in adaptation. As referenced in Table 1.1, many of these challenges are deeply embedded in institutional limitations and bureaucratic constraints. However, an equal number of challenges are based on a lack of market experience with and/or a risk-tolerance for the types of innovation necessary to develop appropriate underwriting criteria for performance and risk.

However, adaptation finance also represents an opportunity to engage new value chains that offer co-benefits for transportation, affordable housing, ecological conservation and public health, among many others. As referenced in Chapter 3, many of the opportunities to fund adaptation are derived from programs that offer only indirect support for co-benefits consistent with adaptation. In this case, the challenge is to understand how sustainability goals might align with adaptation ambitions. Table A.1 provides a range of co-benefits by and between adaptation, resilience and sustainability. What is increasingly well established is that adaptation finance is not simply about finding dedicated funding streams in isolation to fund a project that is designed for a singular purpose of advancing a particular adaptation and/or resilience intervention. To the contrary, the challenge is to finance the incremental and marginal costs associated with adding adaptation and resilience elements (or strategies) to an existing asset or investment. That is to say that people do not build reinforced bridges to withstand increased flows from flash floods; rather they build bridges as a means to facilitate a transportation system.

With this marginal cost framing in mind, there are several challenges that will shape the development and maturity of adaptation finance in the immediate future. The first challenge relates to flexibility. For example, existing funding sources for post-disaster response are primarily oriented towards immediate recovery with some potential for resilience, but they are generally not flexible enough to adequately accommodate long-term adaptation investments. Going forward, the challenge for developing adaptation finance programs and conduits is to ensure that local experimentation is supported in a manner that allows for the development of more refined criteria

Table 1.1 Common barriers to adaptation funding and finance

	Barriers	*Explanation*
Institutional	Low public policy priority	Many other competing investments
	Lack of leadership	No incentives for championing investments
	Conflict of interest	Strong local government interest in ignoring climate change
	Disproportionate burden/prior disadvantage	Small and minority communities and businesses disproportionately bear the costs
	Siloed government	Lack of necessary multi-jurisdictional coordination
	Lack of administrative capacity	Lack of capacity and training to seek and manage complex funds
Economic	Disjointed risk structure	Challenge in valuing non-market benefits, as well as lack of a capacity to account for uncertainty, inaction and proper discount rates
	Inability to make the economic case	Growing internal competition and a lack of capacity to increase revenue
	Chronic underfunding	Local funding likely inadequate to address regional problems
	Inappropriate funding scale	Misalignment between funding cycles and changing conditions
	Discontinuous funding	Misalignment between funding cycles and changing conditions framing investments, as exemplified through pre- and post-disaster funding
	Aversion to innovation	Innovation is too risky and difficult to scale without necessary experiments to develop criteria for performance
Financial	Funding bias	Biases towards planning large discrete projects and not for implementation or broader programmatic adaptations
	Lack of knowledge about funding	Funding sources are siloed and required complex aggregations that are circumstantially defined
	Restrictions and eligibility requirements	Patchwork of funding sources that lack internal consistency

Source: Adapted from Moser et al. (2018).

for underwriting and project assessment. Therefore, it will be important that financial conduits are flexible enough to manage a variety of debt and equity products. This may mean overlooking or cross-subsidizing transaction costs and credit risks for smaller investments that represent potentially valuable proof-of-concept experiments.

In addition, across an entire portfolio, the challenge will be to provide flexibility between commitments of the funds and the actual deployment of those dollars given the long lead times for planning and permitting. Beyond capital deployment, reporting time horizons and cash-flow stabilization periods will need to be modified to account for the asynchronization that climate change shocks and stresses may place upon certain assets. For instance, quality assurance may be negatively impacted by the utilization of new materials, construction techniques and asset management practices that have limited historical precedent. This is a particularly important consideration for financial underwriting, as yield attainment and maintenance on many assets is highly dependent on the performance of the asset in periods extending well beyond initial stabilization.

An additional challenge relates to coordination and guidance challenges from federal, state and local agencies and stakeholders. Whether it is coordinating consistent discount rates for pooled multi-agency funds (e.g., conduits) or synchronizing design standards that are based by some measure on precautionary principles for addressing the upper boundaries of climate scenarios (i.e., worst-case scenarios), there are plenty of gaps in matters of policy that will operate to create uncertainty in planning, execution and performance. Likewise, practices determined by professional societies, standard development organizations and private contracting will also create novel relationships that operate to create and shift risk among parties in untested ways. For instance, emerging contracts for the professional services of architects and engineers are extending liability for the assessment of long-term climate change impacts on the design and performance of buildings and infrastructure, whether or not such considerations are required by code. Given the wide-ranging implications of climate change impacts across all sectors of the economy, it is likely that private market behaviors will adapt much faster than public policies can accommodate. Therefore, potential friction between public and private sector goals and values is to be anticipated.

The final challenge relates to the development of markets, products and services. Novel alternative underwriting techniques and data services (i.e., climate services) will need to be developed to ensure adequate transparency for identifying relevant risks, uncertainty and costs. As previously referenced, NPV optimization has a limited value given the qualitative trade-offs associated with the complex negotiation of interests and values in the public and private sectors. Robust decision-making (RDM), cost-effectiveness analysis (CEA), real options analysis (ROA) and portfolio analysis (PA) methodologies will likely need to be developed with specific application to understanding where and when it is best to deploy capital allocations. While these alternative analytical methodologies are beyond the scope of this guide, there are a number of emerging consulting firms, including several in California, that are developing expertise in everything from climate disclosures for public companies to targeted valuation techniques for large portfolios.

What unites all of these techniques and firms is the necessity to rely on a stream of data and intelligence about not only climactic change, but other indicators associated with global change; social behavior (e.g., consumer preferences); adaptation responses and feedbacks; supply chain performance; climate derivatives and insurance pricing; and the pricing and performance of adaptation investments. While there are technology firms that have begun to model and license data streams, there are few, if any, one-stop-shop aggregators of data and products. This field known as "climate services" is growing quickly and is likely to continue to advance in the coming years as low-cost monitoring hardware, satellite capacity, industry awareness and market demand proliferate. One outstanding challenge going forward for climate services is the capacity to authenticate data to a point in time that counter-parties can agree on for purposes of risk assessment, risk transfer, event triggering, valuation and pricing. For instance, if two different investors are using two or more different data sources to measure a triggering event (i.e., payout or default), then how does one objectively resolve not only the optimal data source but the validity of that source given the variability and manipulability of unsecured – often open source – models and data sources? Current research is exploring the possibility of using blockchain technology to standardize and authenticate such data for purposes of optimizing transactional efficiency.

2 Climate adaptation and asset management

The enumeration of climate change impacts on various sectors is widely established through the ongoing assessment process of the California Climate Change Assessments (CNRA 2018b) and the National Climate Assessments (USGCRP 2018b). This chapter seeks to advance a process by which local governments and asset managers can begin to synthesize climate assessments with asset management assets. This chapter is not intended to be exhaustive in identifying the connection between impacts, stresses, shocks and hazards by asset class. Rather it is intended to frame a set of inquiries and processes that begin to internalize types of risk and opportunity that have historically fallen outside of the purview of conventional practices in asset and portfolio management.

Asset management processes

For purposes of this chapter, the term "asset" is primarily oriented towards capital assets that have some degree of climate sensitivity. However, upon closer inspection, nearly all assets have some measure of climate sensitivity (Mercer 2011). Even cash has some degree of climate sensitivity. For instance, cash availability in post-disaster recovery scenarios is a significant challenge and current federal policy for alleviating such constraints is primarily limited to items such as homeland security and debris clearance. Sovereign fixed income, investment grade credit, developed and emerging market equities and private equity all have some degree of short- and long-term exposure and sensitivity to climate change (BlackRock 2016). In this regard, it is not only direct economic output or physical asset impairment that affects these asset classes; it is also a function of consumer and investor sentiment about how climate is, could, would or will lead to devaluation (Coburn et al. 2015). However, for most local governments, these are considerations best accommodated by third-party managers.

For local governments, developing an understanding of the relationship between the economics of climate adaptation and asset management is central to not only evaluating risk but also the opportunity associated with adaptation investment. This can be undertaken at all levels of governance from public works and solid waste to public safety and human resources. While the predominant consideration is for physical and material assets and engineered systems, similar methodologies may also be applied to understanding vulnerabilities associated with social capital. This may include human resources, operations and communications processes, and knowledge transfer processes. This may go well beyond business continuity assessments to include post-disaster considerations, such as mobility and identification and verification resources, as well as an understanding of the stresses that climate change impacts may have on the availability and affordability of housing and healthcare services.

The more immediate challenge is to develop asset management processes that can assess the potential impact of climate change on assets; develop criteria for assessing impacts and interventions; and, execute and monitor investment strategies. As per Figure 2.1, the first inquiry

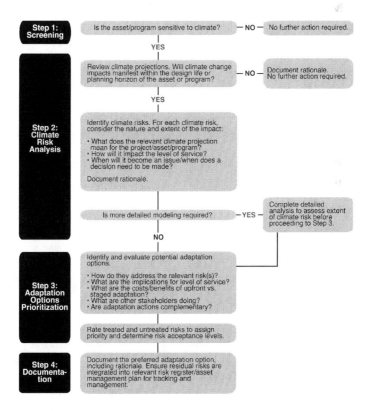

Figure 2.1 Decision tree for climate adaptation asset assessment

Source: Adapted from GSA (2015).

should revolve around whether the asset (or program) is sensitive to climate change impacts. This may require consultation with third-party subject matter experts, as well as resource guides from a range of sector-specific sources, including the U.S. National Institute of Standards and Technology (NIST) (2015a, b), the American Society of Civil Engineers (ASCE) (Olsen 2015), Transportation Research Board (TRB) (2018), American Water Works Association (2018), U.S. Department of Energy (2016), U.S. Department of Homeland Security (communications) (2015) and Enterprise Community Partners (buildings) (2015). Measurement methodologies for evaluating physical climate risk are emerging, but the publication *Bridging the Adaptation Gap: Approaches to Measurement of Physical Climate Risk and Examples of Investment in Climate Adaptation and Resilience* provides a useful survey of various approaches (GARI 2016). If a particular asset is deemed not to have a climate sensitivity, then the process of evaluation and the associated findings should be documented in order to facilitate future audits and reassessments.

Assuming that a sensitivity does exist, the next step is to evaluate existing climate models and projects to evaluate whether such potential impacts may manifest within the design life and/or planning horizon of the asset or program. As previously referenced, the companion guide, *Planning and Investing for a Resilient California: A Guidebook for State Agencies* (OPR 2017), provides a range of resources for models and climate services that are most appropriate for California's varied geographies, ecologies and climates. If upon further investigation it is determined that climate impacts will either manifest beyond the design life of the asset or that such impacts occurring within the design life are not anticipated to impact performance, accelerate aging effects or otherwise have negative implications for an asset, then the analysis should be documented and cataloged to facilitate future audits and reassessments. However, if the findings suggest an adverse impact, then the analysis must evaluate risk – that is a probabilistic (i.e., likelihood) assessment of the negative impact of phenomena. Non-probabilistic assessments are relegated to matters of uncertainty. As per Figure 2.2, each asset can be assessed for its risk based on a likelihood (i.e., probability) relative to its impact (i.e., damage, cost). Furthermore, as referenced by example in Table 2.1, probabilities can be translated into simple heuristics that reference both reoccurring and singular events and return periods. The heuristics of likelihood (e.g., likely, possible, unlikely, etc.) are a useful way to translate complex probabilities that may vary in terms of distribution and variance.

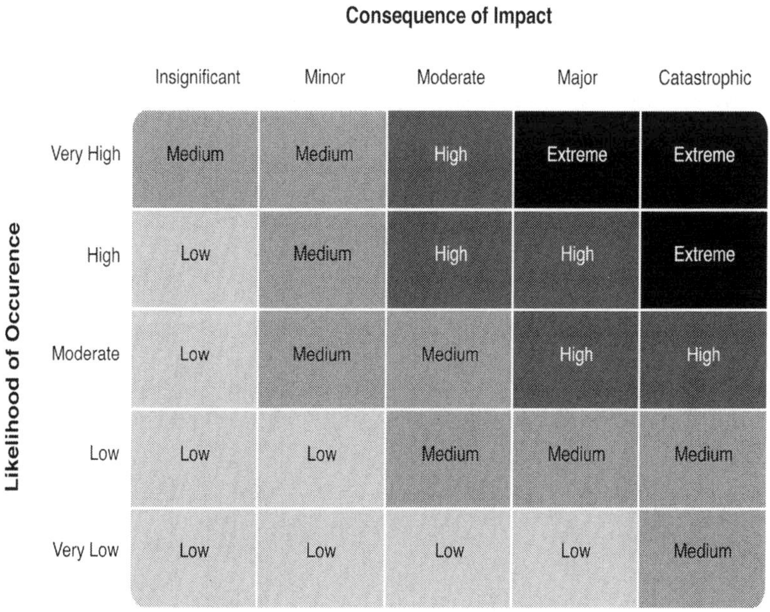

Consequence of Impact

		Insignificant	Minor	Moderate	Major	Catastrophic
	Very High	Medium	Medium	High	Extreme	Extreme
Likelihood of Occurence	**High**	Low	Medium	High	High	Extreme
	Moderate	Low	Medium	Medium	High	High
	Low	Low	Low	Medium	Medium	Medium
	Very Low	Low	Low	Low	Low	Medium

Figure 2.2 Risk prioritization matrix
Source: Adapted from USGSA (2018).

Table 2.1 Example of heuristics for occurrence likelihood

Heuristic	*Recurring events*	*Single events*
Almost certain	Several times a year	More than likely to happen
Likely	About once a year	As likely as not to happen 50/50
Possible	About once in three years	Likelihood less than 50/50
Unlikely	About once in ten years	Likelihood low but not negligible
Rare	Less than once in ten years	Negligible likelihood

Source: Adapted from GSA (2015).

Box 2.1 Climate impacts on reserve accounting

Assuming that there are different probabilities for different impacts that are equally distributed and of known quantities, then in a perfect scenario expected values (EV) could be determined for replacement and/or recovery costs (inclusive of accelerated aging effects) that could otherwise be utilized to increase reserve line-items for capital accounting. For instance, consider the impact that extreme heat may have on a roof of a municipal building. The roof has a useful life of 20 years and the present value (PV) for its replacement cost is $100,000. Reserve accounting would utilize a straight-line amortization (1/20) that would dictate that $5,000 a year be set aside to replace the roof at the end of year 20. However, with extreme heat, there is a 50% chance that the roofing material will have an accelerated useful life, which brings the effective replacement cost up to $120,000. Therefore, the EV from the climate change impact is $10,000, which translates to a $500 annual increase (10%+) in cash outlays to fund the reserve account. See Figure 2.3.

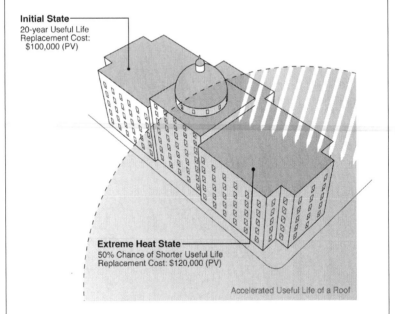

Figure 2.3 Climate impacts on reserve accounting

While the probability of the occurrence and/or intensity of an impact may be understood, that does not mean that the impacts will be consistently or equally distributed by and between similarly situated assets. In systems engineering, resilience can be reduced to an economic consequence that accounts for the time and expense of identifying, diagnosing, resourcing and effectively returning an asset to its pre-perturbation level of service (Hosseini, Barker, & Ramirez-Marquez 2016). Therefore, the levels of performance or service of an asset, as well as the recovery and/or replacement costs, may vary significantly. Close attention should be paid for when investment and management decisions should be made during the life-cycle of an asset. This may require additional detailed analysis that not only stochastically simulates impacts but also the human and organizational responses to those impacts. Current engineering economics research has just recently begun to look at the impacts of climate stresses and shocks and the interactions that long-term stress has on the engineering resilience of a system to recover from a shock. Unfortunately, with imperfect information from which to model, these types of analyses may be grossly incomplete. As a substitute, scenario planning within an organization may identify additional costs associated with diagnosing, resourcing and addressing climate impacts.

Box 2.2 "Fat tails" and climate change

One of the underlying concepts associated with climate change is that a shift in the mean (e.g., temperature) does not necessarily correlate with a corresponding shift in the existing distribution. As a general proposition, referenced in Figure 2.4, a shift in mean is anticipated to flatten the distribution. The implication is that the probability of

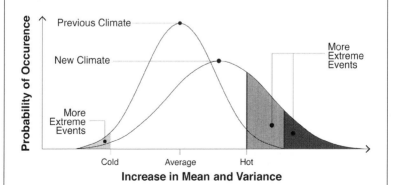

Figure 2.4 Climate change-driven shift in mean and variance

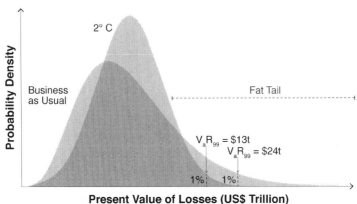

Figure 2.5 Climate change-driven shift in tail risk

Source: Adapted from Fuss (2016).

Table 2.2 Current climate return analysis

Scenario	Return (or R)	Probability (P)	(RxP)	R-ER	Px (R-Expected R)²
Projection	13.00%	79.00%	10.27%	0.00230	0.00000
Optimistic	14.50%	10.00%	1.45%	0.01730	0.00003
Below avg.	10.00%	10.00%	1.00%	−0.02770	0.00008
Poor	5.00%	1.00%	0.05%	−0.07770	0.00006
Expected return [sum of RxP]			12.77%		
Variance (V) [sum of Px(R-Expected R)²]					0.0002
Standard deviation (Std.) [square root of the variance]					0.0131
Coefficient of variation [Std./ER]					0.1025

more extreme events increases. In economic terms, this can be trans-lated to the concept of a "fat tail," which can be practically defined as low-probability, high-impact events at the far end of the distribution – often measured beyond three standard deviations (Nordhaus 2011). As referenced in Figure 2.5, fat tails associated with global warming represent a significant increase in the PV of economic losses. The same concept can be applied to assets in terms of average annual losses and exposure reduction premiums.

Table 2.3 Future climate return analysis

Scenario	Return (or R)	Probability (P)	(RxP)	R-ER	Px (R-Expected R)^2
Projection	13.00%	50.00%	6.50%	0.01905	0.00018
Optimistic	14.50%	1.00%	0.15%	0.03405	0.00001
Below avg.	10.00%	40.00%	4.00%	−0.01095	0.00005
Poor	5.00%	9.00%	0.45%	−0.06095	0.00033
Expected return [sum of RxP]			11.10%		
Variance (V) [sum of Px(R-Expected R)^2]					0.0006
Standard deviation (Std.) [square root of the variance]					0.0240
Coefficient of variation [Std./ER]					0.2162

From an investor's point of view, financial returns may also be impacted by a flattening of a distribution. For instance, climate change impacts may have a direct impact on the performance of an investment or asset. As per below, the return analysis under current climactic conditions is not so risky, with an 89% probability that the investment will perform as projected or better than projected. Under the current climate, the expected return is close to the projected return. Under a future climate scenario where probabilities flatten, the difference in expected returns is only 1.68%. However, the risk measured by the standard deviation and the coefficient of variation more than doubles.

Assuming that the sensitivity of assets can be determined based on appropriate projections and models, the next step is to evaluate the extent to which design and/or management interventions that promote the resilience of an asset (or program) have parity within the underlying risk and/or uncertainty. Does the investment mitigate the impact, or does it merely limit the exposure in terms of reducing recovery times and costs? Chapter 4 will explore the parameters of investment analysis of resilience and/or adaptation investments. However, assuming that the resilience investment is determined to be efficient and effective based on known probabilities and return periods, it is necessary to take the analysis further and ascertain if the resilience investment still makes sense if shifts in mean and variance associated tail risk and "fat tails" undermine or increase the value of the investment.

Box 2.3 Impact-variable adaptation investment analysis

With climate change, extreme precipitation events are anticipated to increase in certain geographies. When designing a stormwater system, engineers are considering adapting the system to accommodate 36" over 12" drain pipes for a low-lying neighborhood containing city-owned public housing. The incremental cost of the adaptation is $10,000,000, including the increased costs of processing more stormwater runoff. The average annual losses (AAL) for the public housing are currently $700,000 based on historic flooding. With the adaptation, the AAL are down to $50,000. The difference between the pre- and post-adaptation AAL (i.e., avoided losses) are calculated as net savings and treated as a fictional cash-flow for purposes of investment analysis. With a discount rate equaling the general obligation bond rate of 3% and the recovery costs increasing every year by 3%, the adaptation pays for itself in the 20th year with a positive NPV of approximately $2,600,000. However, at the end of the decade (year 11), the AAL without the adaptation is anticipated to double as a reflection of increased risks associated with extreme precipitation events. Given this shift in probability, the payback period is reduced by five years with a near equivalent positive NPV. See Figure 2.6.

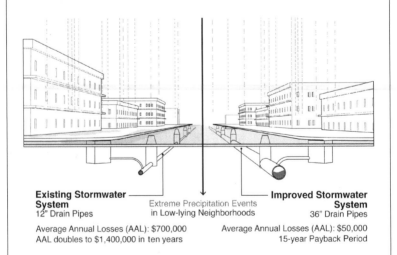

Existing Stormwater ———————————— Improved Stormwater
System   **System**
12" Drain Pipes   Extreme Precipitation Events   36" Drain Pipes
                    in Low-lying Neighborhoods

Average Annual Losses (AAL): $700,000   Average Annual Losses (AAL): $50,000
AAL doubles to $1,400,000 in ten years   15-year Payback Period

Figure 2.6 Impact-variable adaptation investment analysis

As explored through the example in Box 2.3, increased probabilities for extreme events may actually increase the value of an adaptation investment within the associated capital cycle of the investment. This example highlights another factor in the determination of the feasibility or desirability of an adaptation investment – timing. The question is whether the intervention should be made now; not at all; phased over time; or, delayed until such point in time as there is adequate information to support the decision. While beyond the scope of this guide, this later proposition has been the object of emerging research in real options analysis (ROA) (Sturm, Goldstein, Huntington, & Douglas 2017). To make sense of these various decision points, it is useful to conceptualize timing within the context of various adaptation strategies.

Adaptation strategies

Table 2.4 and Table A.2 highlight a range of adaptation strategies that are useful for organizing the value and timing of various investments (Hallegatte 2009; Keenan 2015b). Among these strategies, a determination can be

Table 2.4 Example adaptation strategies for flooding and sea level rise

Adaptation measures	No-Regrets strategy	Reversible/ Flexible strategy	Safety Margin strategy	Soft strategy	Reduced Decision Horizon strategy	Positive synergies with Mitigation and Sustainability strategy
Flood proofing an old building	+		+			++
Infrastructure improvements	+		+			++
Restrictive land acquisitions	−					
Low-cost flood barriers	+	−	+			
Share risk				++		
Transfer risk				+		
Corporate risk management		++		++		
Lower-quality assets					−	
Evacuation					−	

Source: Adapted from Hallegatte (2009), Keenan (2015b).

++ Option yields benefits with or without climate change and flooding.
 − Option yields benefits if urban flooding, but not with inundation from climate change.
 + Option yields loss without occurrence of climate change or flooding.

made as to whether a benefit will be yielded with or without the occurrence of a hazard and/or a climate change impact. The first strategy is a "No-Regret" strategy. This strategy suggests that a benefit may be yielded, even without the occurrence of a climate change impact. For instance, low-cost flood barriers may operate to advance the resilience of a building even if that building ends up not being inundated from increased flooding from sea level rise within the balance of the building's useful life. A similar strategy is a "Low-Regrets" strategy wherein the costs of the intervention are otherwise marginal enough to be entirely written-off. The second strategy known as a "Reversible/Flexible" strategy reflects on the value of maintaining future potential pathways within the context of imperfect information and where alternative options are defined by relatively high first costs.

The "Safety Margin" strategy allows for investments that advance pre-cautionary principles in the design of additional capacity or excess margins of performance. In other words, it is better to be safe than sorry. For instance, this may be an appropriate strategy for critical facilities, but it may not be an appropriate strategy for non-critical assets with relatively low capital inputs and short useful lives. Under these conditions, increased capital allocations for resilience may be maladaptive because replacement reflects a superior outcome. For instance, a pair of shoes costs $50 and lasts two years. A "resilient" pair of reinforced shoes costs $100 and lasts for three years. That extra year of useful life equals the first cost of the non-resilient pair of shoes and is hence not an optimal investment.

"Soft" strategies are those things that don't require material investments. These may be strategies that reflect a change in behavior, such as a change in operations, management and the utilization of an asset. For instance, reducing the use of an air conditioner to make up for extra operating hours on increasingly hot days may be a soft strategy for maintaining the design life of the asset. Soft strategies may also be synergistic with a "Positive Synergies with Mitigation and Sustainability" strategy. In this case, investments made in adaptation may also operate to contemporaneously advance sustainability or other social, economic or environment goals. For this reason, this strategy is sometimes referenced as a "Win-Win" strategy. In carrying forward the example above, an extension of the design or useful life of an asset means a potential alternative reduction in the rate of consumption. As such, with the need to produce fewer air conditioning units, there is a corresponding reduction in greenhouse gases associated with the production of the units. This may also result in a win-win for users in that it also reduces their utility costs.

The "Reduced Decision Horizon" strategy is very often a compelling strategy for investors who seek a point in time where there are more clear signals of climate change impacts, as well as a better understanding of the costs of those impacts. With this strategy, the idea is to make smaller incremental investments that allow for a delay in the more substantive investment without decreasing the relative marginal utility of the later investment. In this regard, the strategy is simply about buying time. The fundamental questions in this case are two-fold. First, how much and what kind of information does one need to make a decision? Second, will options or options at the price points as they exist today still be available at any given point in time in the future? One always runs the risk that it is either too late to make the investment or the costs (or return profile) of doing so are otherwise constrained.

For instance, if it takes 20 years to construct a sea wall, there is the risk that if one waits too long, then intermediate flooding may complicate construction and/or the increased competition for materials (e.g., concrete) and labor for other adaptation projects increases the costs beyond the desired values associated with the underwriting of the project. In this case, it might be more efficient and effective to simply relocate the population that the sea wall was intended to protect. Of course, as will be discussed, this may be a suboptimal outcome in terms of the community resilience of vulnerable populations who may bear inordinate market and non-market costs associated with relocation. In either event, timing is a critical factor and operating with imperfect information and values is a significant challenge both strategically and tactically.

Developing formal adaptation strategies can be an effective way to be more impactful in tactically managing assets and allocating resources. In returning to the concepts relating to timing, investment decision-making may happen along a variety of points along a project timeline. Figure 2.7 represents one way of conceptualizing investment decision-making in light of the long lead time associated by planning, design and execution. In this case, threshold values for sea level rise have been determined so as to provide an upper boundary in terms of time. From the initial decision point (i.e., go-no-go) through the associated lead time, a variety of uncertainties may either be resolved or discovered. From an asset management point of view, it must be determined what thresholds of a particular asset may be and whether the asset has the adaptive capacity for alternative design iterations (i.e., additional capital inputs) to accommodate alternative impact and performance scenarios. For instance, a port may need to make adaptation investments in elevating its crane infrastructure. Because of complex permitting and contemporaneous adaptation activities within the port, the lead time may present a range of uncertainties as to how high the cranes may need to be. In an event of a rapid ice melt scenario (or the significantly increased probability of such an event), it may not be cost-effective to increase the height of the cranes because a threshold may be crossed wherein the entire port may be inundated within the life of the investment. Therefore, it might be necessary to phase certain investments in an order that optimizes the risk

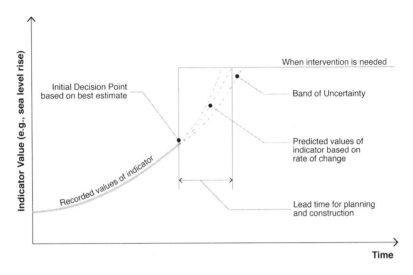

Figure 2.7 Relationship between threshold, lead time and decision points

Source: Adapted from Ranger, Reeder & Lowe (2013).

of a path dependency limiting future options. This example is based on a real equation calculated by a port authority on the East Coast.

As part of the process for considering adaptation within asset management, it is critical to evaluate contemporaneous, complementary and conflicting activities by parties that fall inside and outside of the control of the relevant investment parties. In a best-case scenario, co-benefits may arise between related and unrelated activities that operate to reinforce the values of each (re:focus partners 2015). In a worst-case scenario, excessively delayed investments operate to conflict with each other in a manner that undermines their feasibility and intent. In either event, it is critical to document and rank investment options and alternatives against a matrix of goals, risks and external conditions. This process will be explored in more detail in Chapter 4. Overall, residual risk and uncertainty should be accounted for not only in the analysis of various options and alternatives but also in the performance of the underlying assets that are the objects of such investments.

From asset management to integrated portfolio management

To be effective, asset management activities should be integrated within local government administrative units, as well as across local government portfolios. As referenced in Table 2.5, there are a range of opportunities to connect

Table 2.5 Opportunities to integrate climate change adaptation to asset management plans and policies

Asset management component	Component key climate adaptation components	Opportunity to integrate climate change adaptation
Asset management policy and strategy	Has the organization considered climate change in asset management goals, policies and/or plans?	Incorporate climate change considerations into asset management goals and policies: these could be general statements concerning adequate attention of potential issues or targeted statements concerning specified types of climate risk (e.g., heat waves, flooding, etc.).
Integration strategy with asset management implementation plan	Has the organization mapped areas vulnerable to projected climate risks? Has the organization inventoried critical assets, created risk profiles and developed risk mitigation strategies?	Identify vulnerability of infrastructure assets in areas susceptible to climate change impacts. Inventory critical assets and identify and implement appropriate adaptation strategies (e.g., updated design guidelines, etc.) for these assets or asset classes. These strategies should be mapped to the appropriate administrative unit that will oversee the life-cycle management activities of that asset or asset class.
Key asset management activities	Has the organization considered adaptation strategies at the enterprise, asset class or life-cycle asset management planning level?	Required adaptation strategies in the near term should be identified. Key asset management activities required within the next year can be based on condition assessments where preventative maintenance is warranted to avoid exacerbation of wear and tear or damage due to anticipated climate impacts. It can also involve reactive maintenance activities due to an extreme weather event.
Financial requirements	Has the organization incorporated climate risk mitigation strategies into its short- and long-range plans? Capital and/or OM&R budgeting process?	Costs associated with the key asset management activities (e.g., replacement parts, retrofits, labor, etc.) identified above should be estimated and incorporated into the organization's capital improvement plan and/or operations and maintenance budgets.

(Continued)

Table 2.5 (Continued)

Asset management component	Component key climate adaptation components	Opportunity to integrate climate change adaptation
Continuous improvements	Has the organization begun monitoring asset conditions in conjunction with climate change indicators to determine if/how climate change affects performance?	Monitor asset conditions in conjunction with climate-related conditions (e.g., temperature, precipitation, winds, etc.) to determine how it affects performance; incorporate risk appraisal into performance modeling and assessment; flag highly vulnerable assets. Monitor asset management system to ensure effective response to climate change; possible use of climate-related performance measures or thresholds to identify when an asset has reached a critical level. Revisit life-cycle management plans for assets as appropriate based on performance monitoring.

Source: Amekudzi, Crane, Springstead, Rose, and Batac (2013).

asset and enterprise level activities that allow for the management of climate change risks and opportunities. This requires the development of a broader set of priorities that link intelligence, planning and resources. Under optimal scenarios, local government agencies will link enterprise, asset-level and life-cycle asset management activities to find synergies in information, risk-thresholds and risk-tolerance. As previously referenced, these activities are likely to be compulsory in the future for many local governments in order to maintain an optimal credit rating in the face of inordinate climate exposure.

Part of this process should include an inventorying of assets and the associated climate risks assigned to those assets, as well as what adaptation, resilience and hazard mitigation options may exist. Without this information, local governments will have incomplete information and will continue to invest in Reduced Decision Horizon strategies that independently represent increased risk in light of the observed acceleration of climate change. This requires a detailed accounting and mapping of those assets and corresponding risks that offer a comprehensive view of the portfolio in order to determine what assets are critical and what assets should be prioritized. As listed in Table 2.6, this may require a re-classification of what assets are critical and what assets provide varying degrees of operational enhancement.

Table 2.6 Lifecycle priority definitions for public assets

	Lifecycle/Priority	Definitions
1	*Life safety-critical*	Equipment and capital projects with this priority perform a function(s) that when faulty could cause injury or death.
2	*Operation critical*	Equipment and capital projects with this priority perform a function(s) that when faulty directly impacts our ability to provide revenue service.
3	*Operation support*	Equipment and capital projects with this priority perform a function(s) that when faulty could, over time, have an impact on revenue service operations.
4	*Operation enhancement*	Equipment and capital projects with this priority perform a function(s) that serves to plan and/or enhance revenue service operations.
5	*Operation expansion*	Equipment and capital projects with this priority perform a function(s) that serves to plan and/or expand revenue service operations.
6	*Decommissioned*	Equipment with this priority has been taken out of use/service.
7	*Salvage*	Equipment with this priority is no longer in use and is awaiting salvage.

Source: Amekudzi et al. (2013).

The intersection of portfolio and asset management will require the development of climate-related indicators to support ongoing monitoring and intelligence. This allows for not only a calibration of emergent adaptation strategies but also for the opportunity to update life-cycle management plans. For instance, extreme events and/or ongoing stress may necessitate not only reactive investments but also proactive investments such as preventative maintenance. Particularly in cases of accelerated capital investments, it is critical that this ongoing intelligence be incorporated into long-term capital plans. In many ways, this is the essence of adaptive management – learning and recalibration. Synching asset management and capital planning is important for allocating adequate resources for OM&R, retrofits and fixed labor obligations. It is also critical from a strategic and portfolio point of view because ongoing intelligence about life-cycling and OM&R can help calibrate a more comprehensive understanding of long-term investments beyond the biasing associated with framing first costs. For instance, communities often seek to fund large flood barriers without fully accounting for the ongoing OM&R obligations that can easily register as annual costs equal to 2–3% of first costs.

Emerging barriers to execution

There are some practical and immediate barriers to coordinating asset management plans that can support the adaptation strategies and capital plans. As previously discussed, the most immediate challenge relates to who pays and who benefits from such investments. From a broader public good, this is addressed, in part, by the utilization of social equity weights and other considerations that seek to preference vulnerable populations, as discussed in Chapter 5. However, these conflicts may also arise almost entirely within a local government's administration. One department may be responsible for the planning, designing and procuring of assets whose adaptation or resilience investments may yield benefits that may or may not be resolved or recognized in their own reporting (Keenan 2016a).

This point reinforces that challenges associated with timing more broadly in the performance of adaptation investments rarely yield absolute and immediate benefits, even with Win-Win strategies, as the nature of such "wins" is often contested among parties with different priorities (de los Reyes, Scholz, & Smith 2017). In addition, the formulation and assessment of adaptation strategies is highly dependent on the timing of impacts and the timing of planned adaptation investments with less than perfect information. The timing problem is often partially abated with a focus on exposure and risk reduction reflected in lower insurance costs or capital requirements (Linnerooth-Bayer & Hochrainer-Stigler 2015). However, this reliance is often overemphasized in light of the frequent repricing of insurance. Even if these benefits were fully accounted for, there is limited to no guidance on capital reserves relating climate risks and liabilities across portfolios.

A related barrier is the current lack of accounting rules and controls for assessing not only climate vulnerability but also more immediate asset impairment (Linnenluecke, Birt, & Griffiths 2015; Stechemesser, Bergmann, & Guenther 2015). It can be argued that this level of transparency for local governments is critical for understanding not only portfolio risks and opportunities, but it is also critical for informing a broader public discourse about the nature of what should be prioritized. Climate adaptation will inevitably lead to economic winners and losers – as there are few Win-Win strategies. The challenge is to provide adequate public accounting and transparency to inform democratic processes that must decide the ultimate viability, feasibility and desirability of adaptation and resilience investments. More precisely, this discourse will likely be framed in the short term around the analytical assumptions that this guide seeks to challenge. With the emergence of bond disclosure litigation in California, there is likely to be increased pressure to advance the accounting and reporting controls necessary to accommodate increased transparency for public disclosures in the future.

To accomplish this array of tasks, it will be necessary to invest in the institutional capacity necessary to provide the analytical rigor required for sound accounting, accurate reporting and well-informed investment decision-making. This requires greater literacy and awareness across a wide variety of portfolios, departments and personnel. Training and education are a critical part of not only ensuring minimal competence but also ensuring that the intelligence is adequately shared and diffused across information domains. The alternative is the proposition that institutional lock-in will reinforce investment path dependencies that will undermine adaptation strategies and limit the capacity to make necessary or critical adaptation investments in the future. This will require coordination from a variety of stakeholders by and between disaster management and capital planning and between public asset managers and private sector auditors, accountants and financiers. For every adaptation investment that local government makes today, it might mean one less adaptation option in the future. The following chapter provides a survey of a variety of funding sources that highlight the opportunities associated with building momentum through the utilization of a diversity of platforms.

3 Funding and financing adaptation

In practice, the words "funding" and "financing" are often used interchangeably. Financing refers to the broader process of combining sources of debt and equity to provide capital for an investment. A "capital stack" is the multi-layered combination of debt and equity existing at any given transactional stage of an investment. Debt can come from bonds, mortgages and a variety of secured and unsecured loans. Equity has conventionally been referenced as the unrestricted contribution of cash into an investment. In light of the increasing complexity of public investment, the concept of equity should arguably be diversified. For instance, in-kind contributions to planning activities, matching funds for grants and collateral efforts that advance a project could all be considered types of equity. Likewise, these may be registered as present, as well as deferred, contributions. A more diverse accounting of the notion of equity is critical for maximizing and leveraging the total public contribution to any given investment.

Many commentators and analysts have observed that the financing component of adaptation investment is pretty straightforward and does not deviate much from existing practices. As this guide argues, the fundamental deviation is in the underwriting assumptions relating to the performance of assets; the timing of the accrual or recognition of benefits; and, the application of probabilistic and non-probabilistic assessments to the utility and expected value functions of assets. Given existing capacities for intelligence and analytical modeling, many of these underlying evaluations are within the grasps of a broad range of actors.

What remains more elusive is the source of funding. As will be discussed in Chapter 4, the accessibility and sustainability of funding is presently a significant challenge that eclipses any real market limitations. In theory, if there is a business case, there will be money to finance an investment. However, this is not always the case for a variety of reasons. The real question is how much return is needed for one to be willing to make such an investment? The price of adaptation investment is the convergence of two phenomena. With more experimentation and evaluation of innovative products and procurement models, the transactional costs should reflect greater efficiencies that make adaptation investment more mainstream. The flipside is that mainstream public finance, notably the municipal bond market, is anticipated to feel increased pressure from a failure to make adaptation investments. As the value of inaction (both positive and negative) is better understood, the convergence of these two forces is likely to result in an increasing sophistication in both public and private market actors. With greater sophistication comes the opportunity to operate at scale. As some have argued, there are tremendous opportunities to conceive of local, regional and state financing conduits that can serve to strategically and efficiently allocate funding (RBD 2017; Keenan 2018).

This chapter provides a variety of references to local, state and federal sources of funding that may facilitate either the financing of an adaptation investment or may independently support a broader adaptation strategy. Many of these funding sources may not be immediately intuitive as to their application or value. But, many of these sources have been utilized and interpreted in creative ways to help fill gaps between needs and resources. The strategic combination of many of these funds and the requisite coordination to administer and manage the combinations is beyond the scope of this guide. Yet, the opportunities are nearly endless, particularly with disaster risk management and sustainability funding sources that offer a variety of co-benefits. As highlighted in Figure 3.1, the landscape for sources of adaptation funding and financing is broad. While some funding sources have restrictions on how, when and where the money can be spent, there is still a great deal of room for innovative adaptation. The following sections of this chapter highlight an equally diverse range of opportunities.

Local government

Bonds

General Obligation (GO) Bonds: GO Bonds are commonly used to finance public infrastructure and may be backed by revenues generated from local property tax or fees. Local government GO Bond issuances are authorized by a two-thirds supermajority vote of voters in the issuing jurisdiction (Art. XIII, California Constitution). In addition to issuing GO Bonds for dedicated adaptation-focused projects, local governments may incorporate adaptation goals in broader GO Bond financed projects. For example, the City of Berkeley issued $100 million of GO Bonds to repair and upgrade the city's aging infrastructure. In phase one of implementation, the city included adaptation strategies like green infrastructure and bioswale installation for

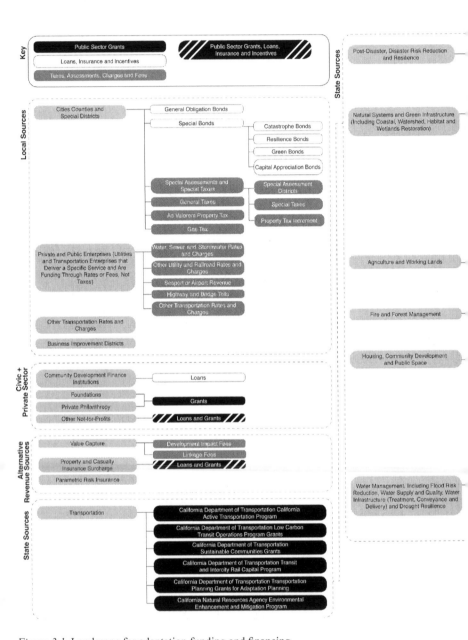

Figure 3.1 Landscape for adaptation funding and financing

- California Infrastructure and Economic Development Bank and California Lending for Energy and Environmental Needs (CLEEN) Center
- IBank Infrastructure State Revolving Fund Program
- Emergency Management Performance Grants
- Greenhouse Gas Reduction Fund
- California Coastal Conservancy Climate Ready Program
- California Coastal Conservancy Proposition 1 Grants
- California Department of Fish and Wildlife Grants
- California Ocean Protection Council Proposition 1 Grants
- California Ocean Protection Council Proposition 84 Grants
- California Senate Bill California Drought, Water, Parks, Climate, Coastal Protection and Outdoor Access for All Act of 2018
- California Wildlife Conservation Board Climate Adaptation and Resiliency Program
- Department of Parks and Recreation Habitat Conservation Fund Grants
- California Department of Food and Agriculture Healthy Soils Program
- California Department of Food and Agriculture State Water Efficiency and Enhancement Program
- Sustainable Agricultural Lands Conservation Program
- California Department of Forestry and Fire Protection California Climate Investments Forest Health Grant Program
- California Department of Forestry and Fire Protection California Forest Improvement Program
- California Department of Forestry and Fire Protection Urban and Community Forestry Grants
- California Natural Resources Agency Urban Greening Program
- California State Parks' Office of Grants and Local Services Program
- California Department of Housing and Community Development Housing Related Parks Program
- Strategic Growth Council Affordable Housing and Sustainable Communities Program
- Strategic Growth Council Transformative Climate Communities Program
- Department of Water Resources Integrated Regional Water Management Grants
- Other Department of Water Resources Grants
- State Water Resources Control Board Clean Water State Revolving Loan Fund

Federal Sources

- Post-Disaster, Disaster Risk Reduction and Resilience
  - Federal Emergency Management Agency Hazard Mitigation Grants
  - Federal Emergency Management Agency Pre-Disaster Mitigation Program
  - Federal Emergency Management Agency Flood Mitigation Assistance Program
  - Department of Homeland Security Regional Resilience Assessment Program
  - Community Development Block Grant Disaster Recovery Program
- Natural Infrastructure (Including Coastal, Watershed, Habitat and Wetlands Projects)
  - National Oceanographic and Atmospheric Agency Coastal Resilience Grants
  - National Oceanographic and Atmospheric Agency Office of Coastal Management Grants
  - United States Department of Agriculture Agricultural Conservation Easement Program
  - United States Fish and Wildlife Service Grants
- Agriculture and Working Lands
  - United States Department of Agriculture Natural Resources Conservation Service
  - United States Department of Agriculture Risk Management Agency Crop Insurance
- Fire and Forest Management
  - United States Forest Service Grants
- Housing, Community Development and Public Space
  - Department of Energy Property Assessed Clean Energy Program
  - Environmental Protection Agency Smart Growth Grants
  - Federal Historic Preservation Tax Incentives
  - United States Department of Housing and Urban Development Community Development Block Grants
- Water Management, Including Flood Risk Reduction, Water Supply and Quality and Drought Resilience
  - Bureau of Reclamation WaterSMART Water and Energy Efficiency Grants
  - United States Environmental Protection Agency Water Infrastructure and Resiliency Finance Center
  - Other United States Environmental Protection Agency Grants
  - United States Army Corps of Engineers Continuing Authorities Program
  - United States Army Corps of Engineers Planning Studies
  - Clean Water State Revolving Fund
- Transportation
  - United States Department of Transportation Build America Bureau
  - United States Department of Transportation Better Utilizing Investments to Leverage Development Grants
  - Federal Transit Administration Grants
- Public Health
  - Centers for Disease Control and Prevention Climate Ready States and Cities Initiative

stormwater management in coordination with the city's Resilience Strategy (City of Berkeley 2016).

Catastrophe (Cat) Bonds: Local governments may serve as a sponsor of Cat Bonds to help insure against damages and fund recovery efforts in the case of a natural disaster. Cat Bonds are "triggered" when a pre-defined event occurs or a set threshold is observed. In this regard, it operates as a type of indemnity. These events or thresholds may be defined by disaster type, severity or amount of damage, or some other metric or composite index. For instance, Cat Bonds may be triggered by wind speed, seismic rating, flood level or rainfall amount. However, sponsors should be careful when specifying the triggering event and the method for observing or measuring the event. Recent experience in Mexico has suggested that scientific observations, in this case relating to wind speed and atmospheric pressure, may open the door to a denial of a triggering event (Blackman, Maidenberg, & O'Regan 2018). As represented in Figure 3.2, investors fund a collateral account and are paid their principal in the event that a triggering event does not happen. The interest on the principal in the collateral account comes from the premiums or coupon payments made by the sponsor.

In the case of a triggering event, the outstanding principal is forgiven, and the insured sponsor may use the balance of the bond funds held in a collateral

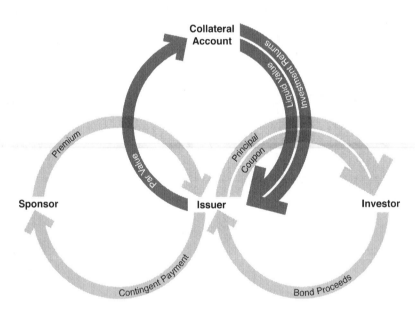

Figure 3.2 Structure of a Catastrophe Bond

Source: Adapted from re:focus partners (2017).

account for disaster recovery efforts. In some cases, the Cat Bond may be triggered, but the amount of actual damages is less than the payout. This serves as a financial windfall to the sponsor. This is different from conventional insurance products where claims adjustment and actual loss are a means to mitigate investor payouts. Therefore, coupon payments and bond yields must account for this transfer of risk. As a high-yield debt instrument, Cat Bonds have high rates of return and short terms, usually reaching maturity in three to five years. They are optimal for low-probability, high-impact events.

Currently in California, Cat Bonds are used to insure against a variety of risks. The California Earthquake Authority currently sponsors Cat Bonds as an insurer-of-last-resort for households at risk of earthquake damage (Swiss Re 2016). While use of Cat Bonds by local governments is more nascent, the New York Metropolitan Transportation Authority through a captive subsidiary insurer issued a Cat Bond in 2013 to insure its facilities against storm events and renewed the bond in 2017 with added earthquake coverage (RMS 2017).

Resilience Bonds: Resilience Bonds are similar to Cat Bonds in that they provide coverage against climate impacts and extreme events, but they also provide financing for adaptation and resilience projects that reduce risk. As represented in Figure 3.3, Resilience Bonds provide financing through rebates to the sponsor local government that utilizes the rebates to make investments that reduce exposure and risk that are reflected in the reduced investment risk to investors and reduced premiums from the sponsor. For example, Resilience Bonds may insure against wind storm event damage while providing financing for property-level upgrades to high-performance roofs and windows (re:focus partners 2017). There are few variations of Resilience Bonds that combine multi-party co-sponsors that pool Cat Bond coverage to "share premiums based on their anticipated risk reductions and dedicated proportionate allocations of their rebate to project implementation or cost-recovery" (id., p. 8).

The major challenge is to link risk with risk reduction strategies and investments that have some modeled parity with enough resolution to justify the rebates. Unfortunately, flooding is considered by many analysts to be too fluid of a risk. In addition, there is a major challenge in operating at a scale large enough to justify any meaningful rebates. For this reason, some analysts have reflected that Resilience Bonds might be more appropriate for national and not sub-national sponsors. In this regard, Resilience Bonds might be appropriate for sponsor organizations that have a wide portfolio of assets. It will likely be a number of years before the market is comfortable with utilizing Resilience Bonds. However, with better intelligence and modeling of climate change impacts, it is anticipated that a better understanding of the probabilistic relationships between risk and risk reduction interventions will help drive the utilization of this instrument.

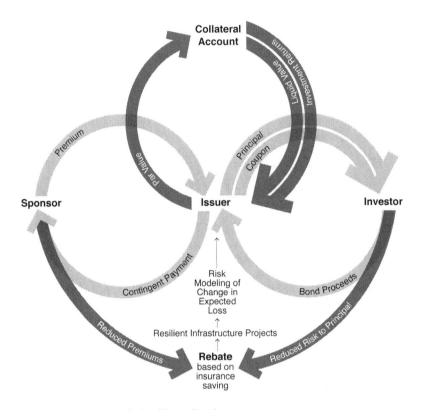

Figure 3.3 Structure of a Resilience Bond

Source: Adapted from re:focus partners (2017).

While Resilience Bonds are not yet in wide use by local governments, California stakeholders are exploring their potential. The City of Los Angeles has been advancing a seismic resilience bond to help finance investments to its water infrastructure (NIBS 2015). Blue Forest Conservation (BFC), a California-based public benefit company, is working with partners at the World Resources Institute, U.S. Forest Service, Sierra Nevada Conservancy and the Sierra Nevada Research Institute to bring a forest resilience bond to market. BFC advocates the potential of the resilience bond to leverage private investment to finance forest and watershed restoration to mitigate wildfire and drought risks. Beneficiaries of the restoration work, such as water and electric utilities, recreation companies and public agency landowners would repay investors based on the success of restoration project efforts (Blue Forest Conservation 2017).

Green Bonds: Local governments may issue bonds to fund projects with environmental or climate adaptation benefits. While there are no formal legal standards by which Green Bonds must adhere to in California, the standards promulgated by the International Capital Market Association (ICMA) and the Climate Bonds Initiative (CBI) are widely regarded (ICMA 2017; CBI 2017). However, it should be noted that there are additional costs associated with the certification and monitoring of Green Bonds when certified by such entities. Green Bonds may warrant lower interest rates from institutional investors who have required internal allocations for green investments, but "this may be of little value for [some state and local governments] that enjoy very low interest rates on tax exempt bonds" (Levy 2018, p. 17). In recent years, Green Bonds have been primarily used in California for renewable energy, energy efficiency, low-carbon transportation, sustainable water infrastructure and pollution control (State of California Office of the State Treasurer 2017). By 2015, just 4.1% of Green Bonds had been utilized for climate adaptation, but there is a consolidated effort in California state government to help facilitate more adaptation-specific issuances (id., p. 12). The California Infrastructure and Economic Development Bank has issued more than $1.3 billion in Green Bonds since 2016. These bond proceeds provide low-cost financial assistance to local agencies under the State Water Resources Control Board's Clean Water State Revolving Fund Program for projects and activities under the Federal Clean Water Act and the State Clean Water Act.

Many local governments in California have issued green bonds, some of which have financed projects with adaptation co-benefits. In 2016, the San Francisco Public Utilities Commission issued a wastewater revenue bond certified as a Green Bond by a third-party external reviewer. Additionally, the San Diego Unified School District issued a Green Bond in 2016 to finance school gardens and a locally sourced farm-to-school nutrition program, solar energy, recycling and waste reduction, energy and water conservation, school bus fuel conservation efforts, and more in the district's public schools (San Diego Unified School District 2010). To varying degrees these investments may reinforce community resilience, engineering resilience for energy systems and facilities, and the systematic resilience of the operations of the school district.

Capital Appreciation Bonds (CABs): Local governments may choose to issue CABs in which the bond principle and accumulated interest is repaid in a single balloon payment when the bond reaches maturity. While CABs allow local governments to defer bond repayment, accumulated interest is compounded. Compounded interest results in a higher overall interest payment and greater overall cost to the local government. CABs have not yet been used to finance adaptation projects in California. However, they have been used by California public school districts since 1993. In 2013, California Assembly Bill (AB) 182 placed limitations on CAB use by school districts. The limitations in AB 182 aim to protect school districts from unfavorable bond terms, including regulating total debt service to principal ratios and maturity dates. CABs might be appropriate for emergency adaptation investments where cash supplies are limited in a post-recovery setting. In theory, if the reinvestment rate of what would have been interim payments are higher than the coupon rate, then it might operate to mitigate the increased overall yield payable in a zero-coupon instrument. Likewise, if the reinvested funds are relatively liquid, then it might be a source of cash for interim extreme events. In general, caution should be exercised in evaluating the use of CABs.

Taxes

Special Assessment Districts (SADs): SADs finance public projects by distributing debt repayment across the property owners receiving special benefits from the project. Depending on the jurisdiction, the assessment may require approval by a majority vote of property owners within the district. Special assessments are appropriate when the project delivers direct and special benefits to specific properties, as opposed to general benefits across a community or region more broadly (RBD 2017). Special Assessment Districts are commonly established to finance road, utility and other infrastructure improvements. In California, Special Assessment Districts have been used for adaptation projects through Geological Hazard Abatement Districts (GHADs) to finance projects to increase public safety and protect specific properties from geological hazards like soil erosion (City and County of San Francisco 2017).

GHADs in California are commonly funded through a tax on property owners within the district. According to the California Association of Geological Hazard Abatement Districts (CAGHAD), there are more than 40 GHADs in the state today (CAGHAD 2011). For example, the Santa Paula GHAD was established in 2008 to provide for the prevention and control of coastal erosion and hazards, such as landslides that could affect properties in a new residential development. The CAGHAD's interpretation of the authorizing legislation (California Resources Code § 26507) has argued that GHADs can be utilized to address sea level rise because it represents an underlying "geological hazard" (CAGHAD 2008). In addition, they cited a GHAD's capacity to accumulate:

> a reserve for future maintenance and rehabilitation, [which] can provide the financial resources necessary for potential future expansion

of flood control structures. Further, because of the relative safety of GHAD revenues (GHADs are typically financed through the collection of supplemental tax assessments), GHADs can borrow from lenders or issue bonds with very attractive credit terms.

(id., p. 4)

Business Improvement Districts (BIDs): A BID is a "program of a city under which the city levies an assessment against businesses or property to fund services or improvements that benefit the assessed businesses or property" (Mandell 2017, p. 1). California local governments may form BIDs one of several ways through a local ordinance, which may or may not require the consent of local business owners (Parking and Business Improvement Area Law of 1989 [California Streets and Highway Code §36500 et seq.]; Property and Business Improvement District Law of 1994 [California Streets and Highway Code §36600 et seq.]). Unlike many other states, BIDs in California are managed by the local government and are not a separate entity. Assessments paid by property owners within the BID may fund adaptation projects within the district, as long as those benefits accrue to the benefit of those businesses and properties being assessed. However, adaptation projects may help achieve local and regional adaptation goals, as local governments may coordinate BID districts across jurisdictional lines with the consent of participating jurisdictions. For example, a BID may use assessment revenues to finance stormwater management improvements or greening to reduce building energy consumption and urban heat islands within the district. These types of improvements often offer co-benefits associated with greater amenities and accessibility for retail and commercial patrons.

Special taxes: As with SADs, special taxes generate dedicated revenue to finance a project by distributing debt repayment across those receiving benefits from the project. However, special taxes are distinct from assessment districts in that they may be used to finance projects with general benefits across a defined geographic region, as opposed to direct and special benefits to specific properties. Therefore, special taxes allow local jurisdictions more flexibility in capturing project benefits. In California, special taxes require a two-thirds majority vote by voters in the district and may be used to secure revenue bonds for project financing (Art. XIII A, §1–7, California Constitution; LAO 2014).

One type of special tax applies to property owners within a defined geographic region receiving benefits from the project. This type of special tax is collected by a Community Facilities District (CFDs) to finance the project. CFDs may determine the special tax formula based on specific community needs and characteristics. Special taxes may be levied on property owners, but are distinct from *ad valorem* property taxes, which may capture up to 1% of assessed property value. Special taxes levied on property that are not *ad valorem* taxes are typically called parcel taxes. Other types of special taxes that may be used to generate dedicated revenue to finance adaptation projects with general benefits include sales taxes, hotel and motel taxes, utility taxes and business license taxes, among others. These and other special taxes may be collected by a city, county or special district or may span multiple jurisdictions.

Local governments in California have levied special taxes for projects with adaptation benefits. For example, County of Los Angeles voters approved Measure A in November 2016, authorizing the County Regional Park and Open Space District to levy a 1.5 cent per square foot parcel tax on all county development to protect, enhance and maintain the area's open spaces, habitats, water bodies and urban tree canopy. Measure A will provide adaptation co-benefits by reducing the heat island effect, protecting local water supply and quality and improving watershed health and habitat, among other benefits (LAC 2018). While this example primarily focuses on sustainability interventions, it highlights the potential for co-benefits between adaptation and sustainability – even though those benefits may accrue to different populations at different times.

Ad valorem **property tax:** Local jurisdictions may use *ad valorem* property taxes, often simply called "property taxes," to finance projects by capturing a percent of the assessed property value within a defined district. Revenue generated by *ad valorem* property taxes is used to finance local government GO Bonds. Local *ad valorem* property taxes are authorized by majority vote of the jurisdiction's electorate and are collected on top of the base 1% state property tax. In California, *ad valorem* property taxes may not exceed 1% of assessed property value (Art. XIII A, §1, California Constitution). Local governments have traditionally used *ad valorem* property tax revenue to finance substantial and long-term projects. *Ad valorem* property tax revenues may service bond debt utilized to invest, for example, in stormwater management projects that restore wetland habitat; replace impervious surfaces; protect drinking and wastewater infrastructure; and, elevate homes and commercial properties in floodplains.

Property tax increment: Local jurisdictions may use property tax increment revenue models to capture the increase in assessed property value within a defined district for the purpose of repaying debt utilized to make improvements within the same district. The increment captures the increase in assessed property value above the valuation at the time of implementation. Property tax increment financing districts may also be referred to as tax increment financing (TIF) districts. Historically, economic development agencies in California used TIF to finance public facilities, services and affordable housing development before the state prohibited this use by economic development agencies in 2011. Today, under a revised statutory authority, Infrastructure Finance Districts (IFDs) and Enhanced Infrastructure Finance Districts (EIFDs) use property tax increment revenues to finance infrastructure projects (CALED 2017). Public Financing Authorities (PFAs) may establish and govern EIFDs to capture property tax increment revenue to finance adaptation projects. Local governments and joint powers authorities are authorized to jointly establish PFAs (Amador 2016). Tax increment revenue "for debt financing is limited to the current annual increment amount less a coverage ratio for security. Issuance of bonds by the EIFD requires a [55%] approval by registered voters within the EIFD" (RBD 2017, p. 12).

Some local governments in California are considering the potential of TIF to finance adaptation projects. In recent years, the City of Los Angeles has been studying the use of an EIFD to finance improvements along a corridor of the Los Angeles River, with improvements including ecological restoration and flood control interventions in coordination with U.S. Army Corps of Engineers (USACE) (HR&A 2016). Likewise, the City of San Francisco and San Francisco are considering the potential of an IFD over Port Authority property to finance part of the San Francisco Seawall. SB 1085 (1995) authorized the San Francisco Board of Supervisors to establish an IFD over Port of San Francisco property to finance public improvements through increased city property tax revenue resulting from the improvements. To utilize the IFD for Seawall financing, the Board of Supervisors needs to approve an Infrastructure Financing Plan to dedicate a share of the city's future tax increment revenue from Port projects to the Seawall (City and County of San Francisco 2017).

Outside of California, local governments have used property tax increment revenues to finance infrastructure projects that have adaptation co-benefits. For example, the Chicago Transit Authority (CTA) established TIF districts to finance public transportation infrastructure improvements for the replacement of aging transit tracks, bridges and viaducts so as to increase the system resilience of the CTA (CTA 2016). This CTA project is also notable because the TIF proceeds were utilized as a local match for a Federal Transit Administration (FTA) grant and a Congestion Mitigation and Air Quality Improvement (CMAQ) grant. This case also highlights one of the limitations of TIF – the district must increase in value.

In the case of Chicago, the districts were capturing value from infill urban development that was likely increasing independent of the transit improvements. With climate change impacts, such as sea level rise, there may be circumstances where districts are decreasing in value faster than improvements can mitigate the underlying deterministic risk. To partially mitigate this problem, it might be incumbent upon jurisdictions to encourage a higher intensity of use and greater densities to facilitate speculation and growth. This strategy may lead to "Climate Gentrification" that operates to alienate the very people that the investments were intended to protect (Keenan, Hill, & Gumber 2018). This trade-off between density, adequate enough to support value capture mechanisms, and Climate Gentrification will likely shape coastal adaptation discourse for many years to come.

General taxes: As opposed to special tax revenue, general tax revenue may not be dedicated to a specific project or used to finance debt. General taxes are approved by a majority vote and support basic service provision by public agencies. California local governments have used general tax revenue to fund adaptation projects as line items in the general fund budget. Additionally, some local governments have provided voter assurance and guidance to local agencies regarding the expenditure of proposed general tax increases through the passage of a separate expenditure plan. In this approach, the general tax increase and the expenditure plan are proposed as independent ballot measures and both require a majority vote by jurisdiction voters (RBD 2017). Overall, the limitations associated with the expenditure of general taxes reinforce the value of incorporating adaptation considerations into asset management and capital planning processes across an entire portfolio. A failure to fully account for climate change will impact bottom-line tax liability and the provision of public services.

Gas tax: Local governments may finance adaptation projects using local gas tax revenue exceeding the base gas tax collected by the state. California gas tax revenue supports transportation infrastructure and services and recently increased from 12 cents to 30 cents per gallon in 2017. Increases in local gas taxes require a two-thirds majority vote (Office of the Controller 2018). Gas tax revenues contribute to the state Highway Users Tax Account to fund transportation-related projects, including those with climate mitigation and adaptation goals. As a general proposition, the scale of investment in transportation and the sector-specific experience in leveraging federal and state funds make transportation projects an ideal umbrella for collateral adaptation-related projects (USDOT 2016). In particular, the U.S. Department of Transportation is comparatively one of the more climate change-sensitive and ambitious federal administrative units, particularly as it relates to engineering resilience. Likewise, Caltrans has made a number of significant analytical and programmatic advancements in climate adaptation, including guidance on vulnerability assessments.

Public and private utilities

Water, sewer and stormwater rates and charges: Under California Proposition 218, all public water agencies and other special districts, such as community service districts, public utility districts and more, may charge rates proportionate to the cost of providing service. Public water agencies that provide a clear case for rate increases proportionate to the cost of service provision must notify their customers of the rate increase; hold a public hearing; and, consider all written protests to the proposed increase. The agency may adopt the increase if a simple majority of the district's ratepayers do not submit written protests (LAO 1996). In the case of a majority written protest, the proposed rate increase must be approved by a two-thirds majority vote by ratepayers or 50% majority vote by property owners to be adopted. California Senate Bill 231 approved this process for stormwater rate increases in 2017. This is a potentially valuable source of revenue for projects designed to manage the increased volumes of stormwater associated not only from the increased probability of rain downfall events but also from the increasingly constrained capacity of watersheds to handle such loads from a combination of sea level rise and soil stability challenges from heat stress on vegetation (Hanak & Lund 2012; AghaKouchak, Ragno, Love, & Moftakhari 2018).

Some California water agencies have adopted a tiered rate structure to cover the cost of service to the highest water users and increase drought resiliency (SWRCB 2018). For example, Moulton Nigel Water District in Orange County adopted a water budget-based rate structure and cost allocation model in which all households are allocated a water budget and households whose use exceeds the budget are charged at a higher rate than households whose use remains within the budget. The agency designates all water rate revenues above the cost of water service to the Water Efficiency Fund to support water supply and efficiency projects (Moulton Niguel Water Agency 2017). This plan represents a variety of co-benefits associated with sustainability, climate mitigation and climate adaptation. It speaks to the sustainability of limited and constrained water resources. Likewise, the reduction in water consumption implicitly reduces energy demand consistent with climate mitigation. Finally, as an institutional adaptation, it imposes a rate system that can be calibrated for future water constraints anticipated with climate change.

One final example of innovation comes from Washington, D.C., which has implemented a Stormwater Retention Credit Trading program that allows projects to develop and sell credits based on green infrastructure development that serves to accommodate stormwater runoff (D.C. Department of Energy & Environment 2018). These types of programs work inversely to reduce a local jurisdiction's capital and O&M obligations for stormwater infrastructure. In some scenarios, jurisdictions themselves may enter the market and gain revenue from their own infrastructure development.

Other utility rates and charges: Electric, natural gas, telecommunications and other utilities may finance adaptation projects related to their property, infrastructure or operations that may have additional impacts for overall community and regional adaptation. In California, privately owned utilities are regulated by the California Public Utilities Commission (CPUC). CPUC approval may be required for adaptation projects and/ or financing whose direct or indirect costs are passed on into a rate base. Utilities may be well positioned to help finance adaptation projects not only because of the potential to finance projects with rate payer revenues, but also because protecting critical gas, electricity, broadband, telecommunications and other infrastructure is critical for long-term shareholder value. Furthermore, utility infrastructure may be located in areas vulnerable to climate impacts like extreme temperatures, wind and stormwater and adaptation investments may improve system efficiencies and reduce operating and life-cycle costs.

Box 3.1 Resilience ESCOs for micro-grid investments

Utilities are increasingly looking for ways to finance resilience investments without passing on those costs to the rate base. It is anticipated that in the future, the development of Energy Service Companies (ESCO) could be developed and modified to help finance resilience upgrades. ESCOs work by making energy efficiency upgrades for third parties. The ESCO provides the capital and often serves as the project manager or contractor for the work. The ESCO's principal and interest payments come solely from the difference between what the third party would have paid and what they currently pay. A similar logic could be applied to resilience investments that offer some measure of savings that could provide a regular cash-flow for an ESCO contract. Currently, a similar arrangement is being developed for use in the development of microgrids on the East Coast. See Figure 3.4.

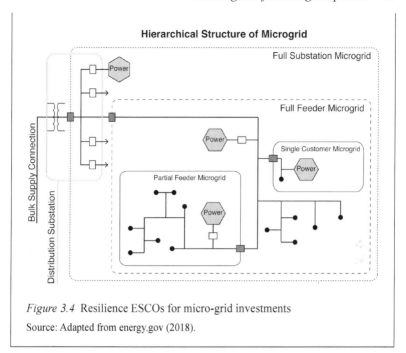

Figure 3.4 Resilience ESCOs for micro-grid investments

Source: Adapted from energy.gov (2018).

Through the Resilience Communities Grant Program, Pacific Gas and Electric (PG&E) provides small grants to local governments and non-profits to promote research and other experimental resilience interventions that advance system and community resilience (PG&E 2018). The interventions must be contextualized within PG&E broader climate change resilience strategies (PG&E 2016). The California Energy Commission's Electric Program Investment Charge (EPIC) program has provided significant funding for resilience research, including the support of distributed energy and micro-grid projects (CEC 2018). Local governments could partner with local research institutions to leverage these resources to evaluate resilience interventions that accommodate their own energy vulnerabilities with climate change.

Seaport or airport revenue: Seaport and airport authorities may collect fees, rates or other types of user charges to service debt or directly fund adaptation projects. While projects undertaken by a seaport and airport authority must provide direct benefit to the respective authority, adaptation investments in seaport or airport property, infrastructure or operations may have broader impacts for local and regional economies. Particularly given the coastal location of seaport and many airport authorities, revenues collected by these agencies may be particularly well suited to financing sea walls and levees. For instance, an airport authority may consider an increase in passengers and airlines fees to finance hard and green infrastructure for flood protection and stormwater drainage, wind barriers or extreme heat and cold adaptation strategies along runways and other critical infrastructure. This is particularly relevant not only because of the low-lying elevation of many runways but also because many runways will need to be lengthened to accommodate longer take-off distances associated with warmer temperatures (Coffel & Horton 2015).

Highway and bridge tolls: Agencies and special districts that manage bridges and highways and collect highway and bridge tolls may finance adaptation projects with toll revenues. In California, toll revenues that have historically been used for bridge and highway operations and maintenance may help finance related adaptation projects. Toll financing requires approval by the California Senate and Assembly and voter approval requirements for toll increases may vary across regional planning or transportation authorities. For example, proposed bridge toll increases by the Metropolitan Transportation Commission (MTC) in the San Francisco Bay Area must first be approved by a majority vote in the state legislature before being put to voters in the commission's nine-county district. Today, a portion of toll revenues collected by the San Francisco Bay Area Toll Authority is used to service bond debt for seismic retrofit projects on seven Bay Area toll bridges (MTC 2017). Bridges are anticipated to endure a range of impacts from climate change ranging from flood-induced scouring to increased traffic demands due to shifting populations on higher elevation settlements (Neumann et al. 2015).

Other transportation rates and charges: Local governments and transportation agencies may consider additional rates and charges to fund adaptation projects. Possible mechanisms might include congestion charging, rideshare fees, license fees or vehicle registration fees within their jurisdiction. Local governments may consider the possible disproportionate impacts of climate change and fees on those without access to public transportation. For instance, the increased occurrence of extreme heat waves may limit the ability of senior citizens to access public transportation. Fees could be utilized to directly subsidize a rideshare program to allow senior citizens to access grocery stores and healthcare services on hot days.

State government[1]

General adaptation and disaster risk mitigation

California Infrastructure and Economic Development Bank (IBank) California Lending for Energy and Environmental Needs (CLEEN) Center: The IBank CLEEN Center provides direct low-cost financing to state and local governmental entities, including public universities, schools and hospitals, for up to 100% of greenhouse gas reduction, water and energy conservation, and environmental preservation projects. IBank financing ranging from $.5 to $30 million is also offered through publicly and privately offered tax-exempt or taxable bonds.

IBank Infrastructure State Revolving Fund (ISRF) program: The ISRF program provides up to 100% financing for up to 30 years to state and local governmental entities and sponsored not-for-profit organizations for infrastructure and economic development projects. Financing amounts range from $50,000 to $25 million and eligible projects may achieve a variety of adaptation co-benefits.

California Office of Emergency Services (OES) Emergency Management Performance Grants: Local governments may be eligible for federal funding administered by California to support general disaster and hazard response planning and preparedness. While climate change hazard mitigation interventions have a limited track-record in this line of funding, the all-hazards approach is broad enough to capture a wide range of activities. In particular, the development of an institutional capacity for communications and intelligence of ongoing hazards is critical for adaptive management of assets and portfolios.

California Climate Investments (CCI): Revenues from California's cap-and-trade program fund the California Climate Investments (CCI) through the CCI. The CCI is managed by the California Air Resources Board (CARB) in terms of accounting, cash-flow management and developing methodologies for quantifying the benefits of associated investments. However, it is the state legislature that is ultimately responsible for the allocation of funds to the state agencies who, in turn, are responsible for allocating funds to specific projects and programs. In this regard, CCI provides local grants through continuous appropriations and budget allocations to state administering agencies. Many CCI-funded California Climate Investments are included in this chapter. Grants support projects that seek to achieve greenhouse gas emission reductions and may include a variety of investments in transit, affordable housing, sustainable communities and high-speed rail (SB 398, 2017). These investments may also achieve co-benefits in terms of hazard mitigation and adaptation. In addition, many grants are specifically underwritten to support disadvantaged and low-income communities. Specific grant programs and funding levels vary with annual state budget appropriations. As referenced in Table A.1, there are a wide variety of potential co-benefits between climate mitigation (or sustainability) and adaptation interventions. The CCI has a co-benefits assessment methodology specific for the determination of adaptation co-benefits for investments made by the CCI (CARB 2018a). The adaptation assessment specifically focuses on project benefits that relate to extreme heat moderation, drought effects moderation, sea level rise and inland flooding, agricultural productivity, species habitat and wildfire. Project categories may range from green infrastructure to urban development and from forest management to shoreline protection. The assessment looks at both positive and negative externalities of a potential project to determine a net-positive co-benefit. While not explicitly addressed in the adaptation assessment, select public health and transportation co-benefits are addressed in separate assessments (CARB 2017a, 2017b, 2018b). Some of the following programs are funded through CCI.

Natural systems and green infrastructure

California Coastal Conservancy (CCC) Climate Ready Program: The Climate Ready Program provides grant funding for nature-based solutions to adapt to climate impacts along California's coast. Climate Ready grants support natural systems approaches in both natural and working lands and human communities and emphasize approaches that achieve co-benefits for communities, the environment and the economy. Some Climate Ready-funded projects have included shoreline planning and design to adapt to rising sea levels, rangeland conservation and planning, carbon sequestration through land acquisition, restoration and conservation, and urban greening and water projects.

CCC Proposition 1 grants: CCC Proposition 1 grants support multi-benefit ecosystem and watershed protection and restoration projects through fish habitat enhancement, wetland restoration, urban greening, sustainability upgrades and more. It is important to remember that viability of ecosystem services can be critical not only for the biophysical performance of ecosystems but also for the fiscal health of local governments that depend on ecosystem services for direct tax revenues, tourism and employment.

California Department of Fish and Wildlife (CDFW) grants: The CDFW supports various grant programs that may increase the ability of habitats and natural systems to adapt to climate change. Local governments may consider CDFW drought response, fish and wildlife management and habitat management grants for project planning and implementation. Some grant programs that may provide adaptation co-benefits include Wetland Restoration for Greenhouse Gas Reduction Program grants, Ecosystem Restoration Program grants, Habitat Conservation Planning Assistance and Acquisition grants, Proposition 1 Restoration grants, and more. CDFW also provides grant funding to support Natural Community Conservation Plans and Habitat Conservation Plans.

California Ocean Protection Council (OPC) Proposition 1 grants: The OPC administers grants using Proposition 1 funding to support marine and coastal areas and water quality. This mission area also includes a variety of ecological resilience activities and investments.

OPC Proposition 84 grants: OPC Proposition 84 grants support adaptive management, marine conservation and research to address ocean acidification, sustainable fisheries and aquaculture, coastal residence and sea level rise adaptation, erosion control and coastal sediment management and more.

California Senate Bill 5 (SB 5) California Drought, Water, Parks, Climate, Coastal Protection and Outdoor Access for All Act of 2018: In 2018, California voters voted to approve a bond measure (Proposition 68), which will provide $400 million for climate adaptation and resiliency projects, in addition to water, community development, open space and other projects that may provide secondary adaptation benefits. As currently planned, funding will become available to local governments in 2019.

California Wildlife Conservation Board (WCB): WCB's Climate Adaptation and Resiliency Program was created in 2017. The program provides local assistance grants to improve public health and the environment, reduce greenhouse gas emissions and strengthen the economy with a focus on disadvantaged and low-income communities using revenues from the cap-and-trade program. WCB also offers other grant programs that may support projects with a variety of adaptation co-benefits.

Department of Parks and Recreation (DPR) Habitat Conservation Fund grants: The DPR Habitat Conservation Fund supports wetland and wildlife protection, acquisition and development with a required 50% local match. With sea level rise, the expansion and contraction of wetlands critical for protecting shoreline settlements will serve as a critical issue for shaping land use patterns.

Agriculture and working lands

California Department of Food and Agriculture (CDFA) Healthy Soils program: CDFA offers grants for implementation of agricultural management practices that may provide adaptation co-benefits. Among other attributes, eligible practices sequester carbon, reduce greenhouse gas emissions and improve soil health. Soil health is a critical component of many green infrastructure systems necessary for the management of water and wastewater.

CDFA State Water Efficiency and Enhancement program: CDFA Water Efficiency and Enhancement grants support projects that reduce on-farm water consumption to achieve greenhouse gas emission reductions and water conservation.

Sustainable Agricultural Lands Conservation (SALC) program: The SALC program provides funding for agricultural conservation easements, agricultural land strategy planning, and other GHG emission reduction projects in the agricultural sector. The SALC program is administered by the Strategic Growth Council in partnership with the California Department of Conservation and the California Natural Resources Agency.

Fire and forest management

California Department of Forestry and Fire Protection (CAL FIRE) California Climate Investments Forest Health grant program: Forest Health grants support projects that help restore forest health to reduce GHG emissions, restore upper watersheds for sustainable water supply, reduce wildfire risk, protect native species and habitat, and more.

CAL FIRE California Forest Improvement Program (CFIP): CFIP grants provide cost-share assistance for forest management planning, tree planting, fish and wildlife habitat improvement and land conservation to achieve productive and stable forests. Forest fires are anticipated to increase in their frequency and intensity in the face of ecological stresses from climate change (Schoennagel et al. 2017).

Housing, community development and public space

CAL FIRE Urban and Community Forestry grants: Urban and Community Forestry grants provide assistance to local agencies and communities to benefit water supply, air quality, stormwater management, energy use, public health and more through tree and vegetation planting and related activities.

California Natural Resources Agency (CNRA) Urban Greening program: Grants administered through the CNRA Urban Greening program use cap-and-trade revenues to support tree planting to sequester and store carbon, reduce energy consumption through strategic tree planting for shade and construct bike and pedestrian pathways to support commuter mode shifts. Urban Greening grants support green infrastructure projects that may provide a variety of adaptation co-benefits, including shading and water storage.

California State Parks' Office of Grants and Local Services (OGALS) Program: OGALS offers grants to support local park and recreation needs, which may provide a variety of adaptation co-benefits, including the support of necessary fire and water buffer zones.

California Department of Housing and Community Development (HCD) Housing-Related Parks program: Administered by HCD, the Housing-Related Parks program supports new park development or existing park improvement through grants to cities and counties. Grants are administered based on the number of newly constructed or substantially rehabilitated affordable housing units to be developed by the applicant agency in the year in which they apply. Housing-Related Parks funding may increase a community's ability to adapt to climate impacts through parks projects with adaptation co-benefits related to flood and stormwater management, heat island cooling, urban agriculture and more. In addition, infrastructure projects, such as flood control projects, may be leveraged to include park amenities.

Strategic Growth Council (SGC) Affordable Housing and Sustainable Communities (AHSC) program: AHSC grants are administered by the SGC and implemented by the California Department of Housing and Community Development (HCD). AHSC grants support projects that achieve greenhouse gas emissions reductions by increasing access to affordable housing, jobs and community destinations through sustainable transportation solutions. Eligible projects include affordable housing development and related housing projects, transportation infrastructure and related transportation projects. Half of grant funds benefit disadvantaged communities. AHSC grants may increase a community's ability to adapt to climate change impacts from co-benefits arising from sustainable, equitable and compact growth and increased access to affordable housing and transit. In particular, redundancy in modes of transportation is a key element of the resilience of transportation systems.

SGC Transformative Climate Communities (TCC) program: Funded through cap-and-trade revenues, the Transformative Climate Communities program provides implementation and planning grants for neighborhood-level plans focused on greenhouse gas emissions reductions that bolster economic, environmental and health benefits in disadvantaged communities. While primarily focused on climate mitigation, there are a variety of resilience and adaptation strategies that are well aligned with this scale of operation, including interventions that address energy and water efficiency that directly increased the robustness of energy and water system resilience. For both the TCC and the AHSC program, climate risk considerations are included within the relevant program guidelines.

Water management, including flood risk reduction,
water supply and quality, water infrastructure
and drought resilience

California Department of Water Resources (DWR) Integrated Regional Water Management grants: Adopted by California voters in 2014, Proposition 1 authorized a state water bond to implement water infrastructure projects listed in regionally adopted Integrated Regional Water Management Plans (IRWMPs). Proposition 1 funding is allocated to IRWMP projects in a competitive grant process. Such projects may include public water system improvements, watershed protection and restoration, integrated water management, water recycling, ecosystem protection, groundwater management, flood management, drought preparedness projects and more.

Other DWR grants: DWR grants support a wide range of water-related projects that may help achieve climate adaption goals. Current grant programs target environmental restoration, flood risk mitigation, sustainable groundwater management, water quality, water supply enhancement and management and water use efficiency.

State Water Resources Control Board (SWRCB) Clean Water State Revolving Loan Fund: Funded by Proposition 1, the SWRCB may provide grants for climate adaptation projects related to watershed and water quality protection and enhancement. This is particularly relevant given the future challenges associated with the provision of water in the face of climate change, including increased competition with agriculture, polluted watersheds from forest fires, increased evapotranspiration and increased reliance on centralized systems, among others (Herman et al. 2018).

Transportation

California Department of Transportation (Caltrans) California Active Transportation program: Caltrans' Division of Local Assistance provides grants to fund alternative and active transportation solutions.

Caltrans Low Carbon Transit Operations Program (LCTOP) grants: LCTOP grants support projects that improve mobility and reduce greenhouse gas emissions, focusing on disadvantaged communities. LCTOP projects may provide adaptation co-benefits, including the provision of greater transportation options for vulnerable populations.

Caltrans Sustainable Communities grants: Sustainable Communities Grants support local and regional planning to help achieve state sustainability goals and may be used to further adaptation goals.

Caltrans Transit and Intercity Rail Capital Program (TIRCP): TIRCP grants support capital improvements to public transit systems to improve operations and reduce state-wide vehicle miles traveled. Infrastructure and operations improvements may provide adaptation co-benefits, including mass transit system resilience. In particular, extreme heat from climate change is anticipated to negatively impact the performance of tracks and energy distribution systems (Chinowsky, Helman, Gulati, Neumann, & Martinich 2017).

Caltrans Transportation Planning Grants for Adaptation Planning: Caltrans Adaptation Planning grants support planning to advance adaptation efforts in the transportation sector with a focus on communities most vulnerable to climate impacts.

California Transportation Commission (CTC): Through SB 1, the CTC State Transit Improvement Program (STIP) provides state grant funding for roadway improvement. One-time grants authorized under this program became available in 2018 and can cover project pre-development costs. Cities, counties and public transit agencies are eligible for funds. Roadways are increasingly under climate stress from increased heat and erosion from increased rain events (Daniel 2017).

CNRA Environmental Enhancement and Mitigation (EEM) program: The EEM program funds projects that mitigate the environmental effects of transportation facilities using revenues collected from the state Highway Users Tax Account, often referred to as the 'gas tax.'

California climate investment maps

The maps on the following pages are static snapshots of investments made through the California Climate Investments program as of November 30, 2017. The maps reflect the public and private recipients of CCI funds, as of November 30, 2017, and do not reflect current or future eligibility requirements. The maps do not represent all of the programs and investments made by the State of California, as covered in this chapter. The metadata and file data for these maps are sourced from the California Climate Investments Project Map, which can be accessed via an online platform: https://webmaps.arb.ca.gov/ccimap/. The intent of these maps is to highlight the programmatic and geographic diversity of investments made throughout the State of California. The maps also highlight the diversity of grantees and beneficiaries, including disadvantaged communities, low-income communities and low-income households. These maps are useful for understanding the sheer scale and impact of the current investment platform, which is insightful for understanding a range of potential adaptation co-benefits in the future. See Figures 3.5–3.12.

Recipients
☐ Public
▨ Private
■ Public and Private

		Public	Private
Natural Resources and Waste Diversion			
●	Organics and Recycling Loans	x	x
▲	Recycled Fiber, Plastic and Glass Program	x	x
■	Organics Grants Program	x	x
○	Urban and Community Forestry	x	
⇵	Wetlands and Watershed Restoration	x	
⌂	Forest Health	x	
⊛	Dairy Digester Research and Development Program		x
Clean Energy and Energy Efficiency			
○	State Water Project: Turbine Replacement	x	
☐	Large Multi-Family Energy Efficiency and Renewables	x	
⊛	State Water Efficiency and Enhancement Program		x
▨	Water-Energy Grant Program		x
⦀	Single-Family Energy Efficiency and Solar PV		x
➘	Single-Family Solar PV		x
Transportation			
☐/○	Low Carbon Transit Operations Program	x	
△	Rural School Bus Pilot Project	x	
⊕	Active Transportation Program	x	
▤/⎯	High Speed Rail	x	
⊙	Car Sharing and Mobility Options Pilot	x	
⁂/⎯	Transit and Intercity Rail Capital Program	x	
○	Public Fleets Increased Incentives Pilot	x	
⊘	Advanced Technology Freight Demonstration Projects	x	
●	Hybrid and Zero-Emission Truck and Bus Voucher Incentive Project	x	x
▬	Clean Vehicle Rebate Project	x	x
●	Zero-Emission Truck and Bus Pilot		x
▨	Enhanced Fleet Modernization Program (EFMP) and Plus-Up Pilot Project		x
Sustainable Communities			
△	Affordable Housing and Sustainable Communities	x	
○	Sustainable Agricultural Lands Conservation	x	

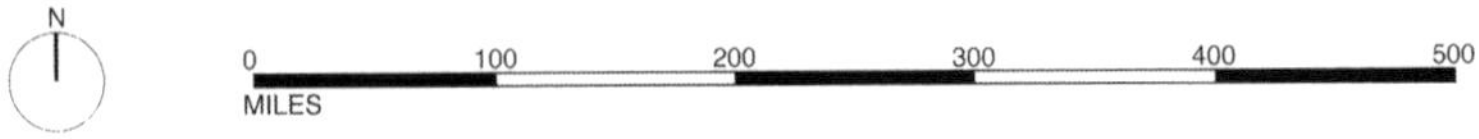

N

0	100	200	300	400	500

MILES

Figure 3.5 California climate investments: legend

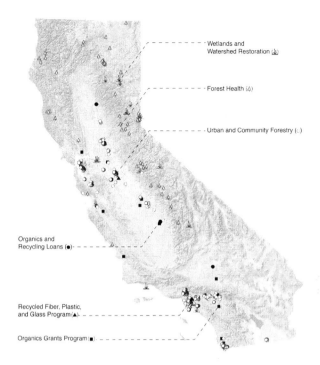

Wetlands and
Watershed Restoration (☵)

Forest Health (△)

Urban and Community Forestry (○)

Organics and
Recycling Loans (●)

Recycled Fiber, Plastic,
and Glass Program (▲)

Organics Grants Program (■)

Figure 3.6 Natural resources and water diversion: public

Organics and Recycling Loans (●)

Recycled Fiber, Plastic,
and Glass Program (▲)

Organics Grants Program (■)

Dairy Digester Research
and Development Program (⊙)

Figure 3.7 Natural resources and water diversion: private

State Water Project:
Turbine Replacement (○)

Large Multi-Family Energy
*Efficiency and Renewables
Part of the Low-Income
Weatherization Program*

Figure 3.8 Clean energy and energy efficiency: public

Water-Energy Grant Program

State Water Efficiency and
Enhancement Program (●)

Single Family Energy
Efficiency and Solar PV
*Part of the Low-Income
Weatherization Program*

Single-Family Solar PV
*Part of the Low-Income
Weatherization Program
Household Level Investment*

Background: Aquifers

Figure 3.9 Clean energy and energy efficiency: private

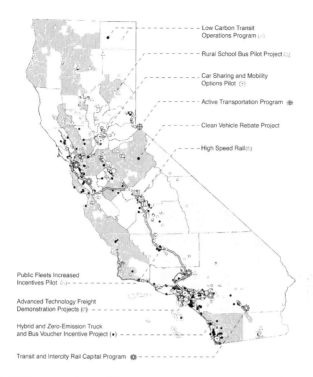

Low Carbon Transit
Operations Program (◦)

Rural School Bus Pilot Project (△)

Car Sharing and Mobility
Options Pilot (∷)

Active Transportation Program ✴

Clean Vehicle Rebate Project

High Speed Rail (⌐)

Public Fleets Increased
Incentives Pilot (◦)

Advanced Technology Freight
Demonstration Projects (○)

Hybrid and Zero-Emission Truck
and Bus Voucher Incentive Project (•)

Transit and Intercity Rail Capital Program ✿

Figure 3.10 Transportation: public

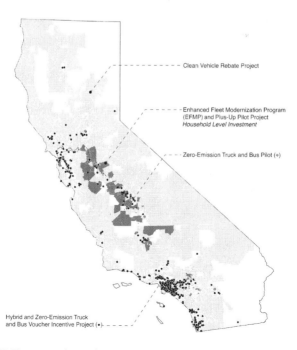

Clean Vehicle Rebate Project

Enhanced Fleet Modernization Program
(EFMP) and Plus-Up Pilot Project
Household Level Investment

Zero-Emission Truck and Bus Pilot (✳)

Hybrid and Zero-Emission Truck
and Bus Voucher Incentive Project (•)

Figure 3.11 Transportation: private

Sustainable Agricultural Lands Conservation (○)

Affordable Housing and
Sustainable Communities (▲)
Background: Population Density

Figure 3.12 Sustainable communities

Federal government[2]

Disaster recovery, disaster risk reduction and resilience

Federal Emergency Management Agency (FEMA) Hazard Mitigation Grants (HMG): Local governments, tribes and eligible not-for-profits may apply for HMG funding in post-disaster areas to implement long-term solutions to reduce or eliminate impacts and future losses from future extreme events and disasters. Potential projects may include increasing the elevation of structures vulnerable to flooding; property acquisition for conversion to open space; flood control projects; wind or extreme temperature retrofits; and, similar projects that advance the resilience of infrastructure of the built environment. This is an area of increasing innovation that many extend to a variety of different novel programmatic options (Stults 2017). Table A.3 provides some potential opportunities for local governments to integrate climate change considerations for their own local hazard mitigation planning (LHMP) into the state hazard mitigation plan (SHMP) consistent with existing FEMA hazard mitigation planning requirements. This is a necessary step for justifying that local investments made with HMG funds are consistent with the SHMP. Consistent with the state adaptation plan (CNRA 2016, 2018a), the 2018 update to the California SHMP contains extensive incorporation of climate change considerations in terms of institutional coordination and technical guidance (OES 2018). Further, the Governor's Office of Emergency Services (OES) has a dedicated Climate Change Working Group that can help guide interagency coordination, including financial planning for projects that "maximize whole community climate readiness and resilience to catastrophic disasters" (Cal OES 2018, p. 663).

FEMA Pre-Disaster Mitigation Program (PDM): PDM grants support projects that implement long-term risk reduction from future hazards and climate events while reducing dependence on future federal disaster recovery assistance. Eligible projects include generator installation at critical facilities, eligible acquisition, elevation and mitigation reconstruction projects, and more.

FEMA Flood Mitigation Assistance Program (FMA): Local governments, tribes and certain not-for-profit organizations may apply for FMA for infrastructure and utility protection, floodwater management, wetland restoration, aquifer storage, water and wastewater management and other related projects.

U.S. Department of Homeland Security (DHS) Regional Resilience Assessment Program (RRAP): The goal of the RRAP is to provide greater understanding of resilient infrastructure needs. RRAP projects are led by DHS and assess critical infrastructure within a designated geographic area to resolve infrastructure and resilience knowledge gaps, inform decision-making, identify opportunities and strategies to improve infrastructure resilience, and establish partnerships between public and private sector stakeholders.

Natural infrastructure

National Oceanographic and Atmospheric Agency (NOAA) Coastal Resilience Grants: Coastal Resilience Grants provide funding for projects that build resilience to extreme events and climate impacts in coastal communities. Grants may support projects that protect coastal property and life, safeguard communities and infrastructure, strengthen the local economy, and conserve and restore coastal environments.

NOAA Office of Coastal Management grants: The NOAA Office of Coastal Management helps coastal communities understand and mobilize the best available information and technology to make informed decisions and increase resiliency to climate impacts.

United States Department of Agriculture (USDA) Agricultural Conservation Easement Program (ACEP): The ACEP provides funding to conserve agricultural land and wetlands and protect their related benefits, many of which may help achieve adaptation goals given the value of ecosystem services associated with agricultural lands and wetlands.

United States Fish and Wildlife Service (USFWS) grants: The USFWS Service provides a range of grants to conserve, protect and restore fish and wildlife habitat in the public interest. Grants are available to local governments and many eligible conservation and restoration projects, such as coastal wetland restoration and preservation, and may provide valuable adaptation co-benefits.

Agriculture and working lands

U.S. Department of Agriculture (USDA) Natural Resources Conservation Service: The USDA Natural Resources Conservation Service provides financial assistance programs to achieve sustainable natural resources management for landowners and agricultural producers. While local governments are not eligible for funds, they may increase public awareness of available funds to help achieve local and regional adaptation goals. Many local governments and their constituents are dependent on the resilience of a local agricultural economy; therefore, programs such as this can be critical for practices that develop a robustness of resources for coping with extreme events.

U.S. Department of Agriculture (USDA) Risk Management Agency Crop Insurance: The USDA offers crop insurance for over 100 crops as a risk management tool for agricultural producers. While insurance is for producers and not local governments, local governments may work to increase public awareness of available insurance programs for growers to help achieve local and regional adaptation goals.

Fire and forest management

United States Forest Service (USFS) grants: USFS offers a variety of grants to support forest health, conservation, economic development and rural communities. As previously referenced, these grants may be combined with state resources to help manage forests in manner that minimizes the destruction and losses associated with an increasing risk of wildfires.

Housing, community development and public space

U.S. Department of Energy (DOE) Property Assessed Clean Energy (PACE) program: The PACE program provides financing for residential and commercial renewable energy and energy efficiency projects that may provide a variety of adaptation co-benefits. In particular, energy resilience may be advanced by autonomous energy generation capacity and the passive performance of buildings. For instance, passively cooled buildings can increase survivability during energy outages for certain vulnerable populations. PACE financing is tied to the property, not the owner, which means that the repayment obligation is transferred to the new owner if the PACE-financed property is sold before repayment. PACE programs are authorized by local government establishing a financing district within which individual landowners may opt in to the PACE program. PACE program participants typically repay the cost of the energy investment over ten to 20 years through property assessments in addition to their property tax bill. Their repayment of the assessment is based on energy savings stemming from the systems financed from either a revolving loan fund or a local revenue bond.

Environmental Protection Agency (EPA) Smart Growth grants: The EPA occasionally offers Smart Growth grants to support community development projects that support human and environmental health. Smart Growth-funded projects may provide a variety of community resilience co-benefits, particularly as they relate to the intersection of housing, transportation and public housing.

Federal Historic Preservation Tax Incentives: Federal Historic Preservation Tax Incentive tax credits may help finance projects with adaptation benefits in designated historic districts or areas. However, local government should consider potential conflicts between adaptation goals and historic building and preservation requirements, such as those conflicts that arise with the utilization of new design assemblies (e.g., saltwater resistant dry-flood-proofing) and modified ingress and egress for increased elevations. Historic preservation boards around the country are challenged with developing climate adaptation plans that allow for modifications to design and materials standards and guidance that can accommodate climate stress while also remaining consistent with the historic character of a building or district (Englander 2015).

U.S. Department of Housing and Urban Development (USHUD) Community Development Block Grants (CDBG): Local governments may use CDBG funding for projects that provide affordable housing and economic development opportunities in low-income and disadvantaged communities. CDBG funds are flexible and may be used to invest in resilience and adaptation. CDBG-funded projects may help achieve adaptation goals through incorporating adaptation strategies in housing, community development and related projects or support implementation of a broader adaption strategy. Specifically, the disaster recovery (DR) variant of the CDBG program has benefited from a number of years of experimentation for funding post-disaster resilience projects and programs. In particular, the National Disaster Resilience Competition (NDRC) offers a valuable compendium of resilience projects and investments that have passed the muster of CDBG-DR program underwriting (USHUD 2018). The experimental projects cover everything from managed retreat to business continuity programming.

Water management, including flood risk reduction, water supply and quality, water infrastructure and drought resilience

U.S. Army Corps of Engineers (USACE) Planning Studies: Local governments may pursue a partnership with the USACE to better understand and plan for area water resource needs and challenges. USACE conducts planning studies to support floodplain management and provide planning assistance to states, local governments and tribes.

U.S. Bureau of Reclamation (USBR) WaterSMART Water and Energy Efficiency Grants: USBR provides 50/50 cost-share funding for projects that improve water conservation and efficiency, water reuse, improve water supply reliability and reduce risk of future water conflict. WaterSMART Water and Energy Efficiency Grants may provide adaptation co-benefits, including reducing exposure to flooding and migrating pollution accelerated by climate change.

U.S. Environmental Protect Agency (EPA) Clean Water State Revolving Loan Fund: This longstanding and popular revolving loan fund offers a below-market mechanism to support a wide range of water quality infrastructure projects. The program offers a variety of different options, including loans, including interest-free loans; repurchase or finance of debt; insurance; and loan guarantees.

U.S. Environmental Protection Agency (EPA) Water Infrastructure and Resiliency Finance Center: The EPA Water Infrastructure and Resiliency Finance Center is a search engine through which local governments may identify federal grants and financing opportunities for drinking water, wastewater and stormwater-related projects. The center helps local governments coordinate funding sources that best leverage available federal funding.

USACE Continuing Authorities Program (CAP): The USACE CAP program may provide funding for the feasibility study and implementation of water and environmental projects related to flood control, aquatic ecosystem restoration, erosion control and prevention, storm damage reduction and more. CAP funding does not require additional project-specific congressional authorization. The feasibility study is federally funded up to $100,000 and any additional costs are shared 50/50 with the project non-federal-agency sponsor. The cost share of implementation is determined by project-specific legislation authorizing a project partnership agreement between the USCE and the non-federal-agency sponsor.

Transportation

U.S. Department of Transportation (USDOT) Build America Bureau: The Build American Bureau provides a variety of credit financing and grants for transportation and infrastructure projects which may include adaptation goals or provide adaptation co-benefits.

USDOT Better Utilizing Investments to Leverage Development (BUILD) grants: Formerly known as Transportation Investment Generating Economic Recovery (TIGER) grants, BUILD grants support transportation infrastructure projects with significant local and regional impact, including road, bridge, transit, rail, port and intermodal projects. Local governments may leverage BUILD-funded transportation and infrastructure projects to achieve adaptation co-benefits, including climate retrofits and system redundancy investments.

USDOT Federal Transit Administration grants: The Federal Transit Administration provides grants to improve public transportation systems. Local governments may use available transit funding to achieve adaptation co-benefits, including providing greater accessibility for vulnerable populations.

Public health

U.S. Center for Disease Control and Prevention (CDC) Climate Ready States and Cities Initiative: The Climate Ready States and Cities Initiative provided grant funding to 16 states and two cities in 2010 to better understand and prepare for the health impacts of climate change. Additional funding is not available at this time, but local governments may monitor the Climate Ready States and Cities Initiative for potential future funding opportunities.

Federal income taxation

Opportunity Zones: Opportunity Zones were created under the *Tax Cuts and Jobs Act of 2017* (P.L. 115–97) to spur investment in economically distressed communities. Among various other provisions, Opportunity

Zones are designated census tracts where investors are able to increase their adjusted tax basis in qualified investments to an amount equal to the fair market value of the investment in the Opportunity Zones on the date that the investment is exchanged or sold, as long as the investor held the investment for a period of ten years (*see generally*, I.R.C. §§ 1400Z – 1, 1400Z – 2). States may nominate zones that are both low-income and contiguous to low-income tracts under certain circumstances (Revenue Procedure 2018–16, 2018–9; I.R.B. 383). In theory, states could focus their nominations in areas that are most vulnerable to climate change and would otherwise benefit from economic development investments. Broad definitions for Opportunity Zone Property could arguably include investments in enterprises and property improvements consistent with localized infrastructure and hazard mitigation projects (26 U.S.C. § 1400Z – 2 (d)(2)(A)).

Civic and private sectors

Community Development Finance Institutions (CDFIs): CDFIs provide financial services to underserved markets and communities with a focus on serving low-income communities through local investments. CDFIs may be community development corporations, banks, credit unions, venture capital funds or loan funds. The United States Department of the Treasury certifies and provides funding to CDFIs. CDFIs can play an important role for building community resilience in disadvantaged communities that are disproportionately vulnerable to climate impacts and extreme events. Furthermore, much of the ongoing community and economic development work of CDFIs is closely tied to strategies that operate to reduce socioeconomic vulnerabilities.

The challenge is to incorporate climate adaptation goals and underwriting processes into existing projects or finance local projects specifically focused on climate adaptation (Donovan 2016). By example, Community Enterprise Partners incorporates adaptation into its ongoing affordable housing development projects by providing technical assistance to developers to incorporate flood and storm resilience elements in building design through the Enterprise Green Communities initiative (Enterprise Community Partners 2015, 2018). Alternatively, the Rural Community Assistance Corporation's Environmental Finance Center provides explicitly climate-focused financing and technical assistance to implement solutions for healthy and adaptive environmental and public health utilities and facilities in low-income and native communities. CDFIs may fund a wide range of projects that impact a community's adaptive capacity to accommodate climate impacts, including but not limited to ecosystem and habitat improvement and restoration, utility and infrastructure upgrades, energy retrofits and credit instruments for small businesses and homeowners affected by climate events. This is an emerging area of interest for CDFIs that are just now exploring how to evaluate either climate changing considerations into their existing portfolios or how to incorporate social equity considerations into ongoing climate-related investments. Chapter 5 explores a few emerging models that speak to this emerging mission area.

Private foundations: Local governments may consider grants by private foundations to fund all or part of an adaptation project. Local governments vary in capacity to seek and administer grants and grants vary in terms of their audit, reporting and general administrative requirements. Grants from private foundations may be particularly well suited for projects that build upon public funds to add further value to existing funding streams or initiatives. Private foundation grants may also be well suited to support projects with greater risk profiles than public funding traditionally supports, such as pilot projects that test new technological or regulatory innovations.

Local governments may consider applying for grants from large adaptation-focused foundation programs, such as those at The Kresge Foundation, Doris Duke Charitable Foundation or the MacArthur Foundation. State-focused foundations like The California Endowment or The William and Flora Hewlett Foundation also have significant philanthropic commitments to climate change. Community foundations may also be particularly interested in supporting community engagement to ensure diverse and/or historically underrepresented stakeholders are included in project planning and development. Private foundations may also be interested in supporting social equity goals to ensure adaptation efforts give the requisite consideration and priority to vulnerable communities.

Private philanthropy: Local governments may consider seeking philanthropic contributions from private individuals. This approach may be reasonable for catalyzing or building early stage support for adaptation investments whose projects have the potential to contribute to the development of community amenities. For example, private donors may be interested to fund park development as part of a larger floodplain or public infrastructure project. As with foundations, private philanthropists may be particularly interested in projects that engage or address the quality of life of vulnerable populations. As such, they may be well suited to provide funding for advancing public engagement that seeks to engage the general public in the design and programing of adaptation projects.

Other non-profits: Many non-profit organizations provide grants for climate adaptation projects or projects with adaptation co-benefits. Local governments may consider funding from national and international climate-focused organizations. Local governments may also consider partnering with California non-profit organizations whose programs provide a variety of adaptation co-benefits in order to coordinate efforts and align public and private interests in the planning of adaptation investments and projects.

Alternative funding models

Development impact fees: Local governments may charge an upfront, one-time fee for new development within a defined geographic or project area to fund projects that offset the negative externalities associated with the development. Impact fees typically finance infrastructure or facilities serving new developments and are generally paid in the course of the permitting processes. Although there is limited experience with climate change, impact fees could be charged to account for impacts associated with increasing exposure of the new development and surrounding geographies to climate change. For instance, water management impact fees could account not only for present impacts but also those impacts that may arise within the useful life of the new projects. This may require infrastructure that has the capacity to accommodate a wider range of design events in the future. Overall, the scale of impact fee revenue is dependent on the level of development within a local government. As such, there may not be parity between incremental units charged as impact fees and the total scale of capital necessary to independently fund such mitigating interventions. However, consistent with the proposition of climate-sensitive stormwater impact fees, local governments may use development impact fee revenue to finance adaptation projects like hard or green infrastructure for flood response and protection, water and sewer system upgrades and urban greening, among other potentially scalable options (Nelson 2018).

Linkage fees: Local governments may require developers to meet certain zoning requirements for the provision of public benefits or pay a linkage fee to provide resources for public provision of those benefits. For instance, many jurisdictions require developers to provide a certain amount of affordable housing in a new residential development or pay a linkage fee to provide public resources for affordable housing construction. Local governments could consider resilience linkage fees to support resilience upgrades to existing infrastructure that is impacted or otherwise has some logical parity with the new development.

Alternative flood insurance models: In theory, local governments may consider resilience and hazard mitigation investments as a means by which to reduce flood insurance premiums and/or deductibles and increase coverage through a corresponding reduction in exposure. However, in practice, there has been little empirical basis from which to justify these reductions relative to the costs or ROI of the investments. As previously referenced, analysts have largely discounted the application of Resilience Bonds for flood mitigation. However, in the future, local governments might be able to pool portfolios to reduce self-insurance exposure.

In California, local governments may also seek to advance further studies for the application of the Community Choice Flood Risk Financing (CCFRF), as an alternative to the National Flood Insurance Program (NFIP) administered by the Federal Emergency Management Agency. CCFRF would offer slightly lower premiums than those offered by NFIP and would otherwise invest the spread between the market rate and the actuarial rate in resilience and risk mitigation projects to further reduce risk exposure (RBD 2017).

Property and casualty insurance surcharge: Similar to a pending proposal in New York, New Jersey and Connecticut, local governments may consider advancing research to impose state surcharges on certain lines of insurance to meet unmet adaptation and resilience needs. Surcharge revenues could be utilized to float bonds in order to leverage an adaptation trust fund that can serve as a financial conduit (Keenan 2018). While insurance surcharges are still an emerging tool for adaptation finance, public agencies have used insurance surcharges to provide consumer and property protection. For example, New York state's Life Insurance Guarantee Fund and Property and Casualty Security Fund protect consumers against insolvency. The states of Kentucky and Mississippi have also used insurance surcharges more broadly to finance projects in the public interest (id.).

Parametric insurance: Local governments may consider parametric risk insurance to finance recovery from climate impacts and extreme events. Parametric risk insurance is based on a model of pre-determined parameters, such as the amount of rainfall in a given period, flood water height or wind speed, wherein the insurance payout is triggered by the occurrence of reaching the pre-determined threshold. Parametric risk insurance is structured to more quickly respond to climate impacts or extreme events than traditional indemnity-based insurance because the pre-determined impact index negates the need for post-disaster damage inspection and evaluation. For instance, payout can be triggered by snow pack on a specific day in the season; the height of a flood stage of a river; or, average sea level on a particular date or range of dates. While the nascent parametric risk market is currently utilized mostly by nation states, parametric risk insurance may be scaled for local governments and a wider range of end users in the future (Howard 2018).

Product innovation in insurance markets is likely to be a critical aspect of public finance in the future. For instance, emerging products are covering the loss of property taxes incidental to the loss of large amounts of properties from forest fires. These types of products essentially serve as credit backstops for accessing lower-cost capital from bonds that are dependent on the revenues from property taxes. As previously referenced, this is particularly important given recent signals by credit rating agencies to more precisely account for climate change risks in credit rating determinations. In the future, it is possible that parametric insurance products could help cover assets and tax rolls from a variety of shocks and stresses associated with climate change.

Notes

1 Website links are provided for each of the state programs in the Appendix.
2 Website links are provided for each of the federal programs in the Appendix.

4   Assessing funding and financing options

The identification of various sources of funding marks an initial stage of development that is complicated by the necessity of evaluating various alternative options. While the criteria for evaluation will be dependent on a variety of qualifications relating to projects, jurisdictions and credit accessibility, this chapter provides some basic criteria that can inform the iterative processes associated with project screening, funding options analysis and the positioning of projects within broader adaptation strategies. Each one of these processes offers insight into the viability, desirability and calibration of different project and investment options. When approached iteratively, they offer an opportunity to right-size investments and to account for the diversity and distribution of costs and benefits. This knowledge will frame critical assumptions for guiding an evaluation of different funding and financing options. However, prior to a more thorough investigation of funding and financing options analysis, it is useful to examine the alignment of the parameters of a project with those requirements inherent in the pricing, availability and terms of various funding and financing options.

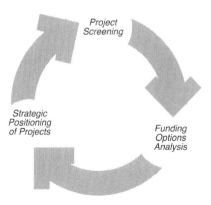

Figure 4.1 Iterative analysis for assessing and weighting financing and funding options

Project screening

An often-overlooked component of the aforementioned options analysis relates to the assumptions made in the initial screening of a project. Because risk and uncertainty are critical elements that will deterministically drive the costs and terms of funding and financing, it is important to have an understanding of how various iterations or manifestations of a project in the earliest stages of development can impact its financial (not necessarily economic) viability. As such, adaptation investments should consider financing options and strategies as early in the planning stages as possible. For instance, projects with irregular phasing and sequencing due to variability of future climate impacts would need to consider the costs associated with credit stand-by commitments, as well as interest rate risks. This phasing may be desirable to optimize ROI as AAL increase with time. As such, project phasing may dictate more or less leverage, depending on the nature of the availability of capital in later phases. This is particularly important as many projects are forced to proceed without firm access to capital in later stages.

From one perspective, there is some measure of quantifiable optimization associated with the development of such strategies, such as is offered in various Bayesian optimization frameworks developed by consulting firms (van der Pol, van Ierland, & Gabbert 2017). However, this guide acknowledges that there are a multitude of external factors, risks and uncertainties associated with adaptation investments. As such, quantitative optimization is useful for informing game playing and scenario planning in the service of strategy development where probability distributions are incomplete (Kruitwagen, Madani, Caldecott, & Workman 2017), but decision-making for project-level screening is arguably best informed through modifications to existing methodological conventions, such as multi-criteria analysis (MCA) and cost-benefit analysis (CBA). This also reflects the reality of operating within state and federal agencies who rely heavily on CBA.

In CBA, the "'best' project is the one that maximizes the expected [NPV] of costs and benefits. Risk aversion can be taken into account through (nonlinear) welfare functions or the explicit introduction of a risk premium" (IPCC 2014, p. 956). There are two fundamental challenges associated with CBA. The first relates to the valuation of non-market benefits (positive and negative) that investments may have on "public health, cultural heritage, environmental quality and ecosystems and distributional [beneficiaries from vulnerable populations]" (id.). However, incremental advancements have been made in alternative valuation techniques (Watkiss et al. 2015), as well as the institutional acceptance of such techniques (USHUD 2015). In situations where alternative investment options have the same annual monetary benefits or "each alternative has the same annual affects, but dollar values cannot be assigned to their benefits," then cost-effectiveness analysis (CEA) is an appropriate alternative (OMB 2016, p. 5). As such, an investment alternative is deemed to be cost-effective above and beyond alternative options if, "on the life-cycle cost analysis of competing alternative, it is determined to have the lowest costs expressed in present value terms for a given amount of benefits." (id.). Therefore, in situations where adaptation investments may yield uncertain monetary benefits but otherwise have some degree of known performance, then CEA may be a reasonable alternative methodology where quantifications of environmental and social benefits come up short.

The second fundamental challenge for CBA, and by extension CEA, relates to selecting the most appropriate discount rate. From a finance point of view, the discount rate can either be conceptualized as a risk premium or a risk-free rate that represents an alternative risk-free (or low-risk) return. In this regard, a "higher discount rate places less value on [benefits] and costs that are further out in time, while a lower discount rate puts more weight on those costs" (OPR 2017, p. 38). But, it may also actually be a direct reflection of the risk and return premium benchmarked to the project through the weighted average cost of capital (WACC) associated with the capital stack. This is where public sector economics and private sector finance diverge. Public sector actors who want to mobilize investment want as low of a discount rate as possible to account for the accrual of lagging benefits with the progression of climate change impacts within the life-cycle of the investment. There is even a line of thought that suggests that negative discount rates are potentially the most ethical route, as it accounts for the multi-generational implications of who causes climate change and who has to pay for it (Fleurbaey & Zuber 2013).

By contrast, private sector actors argue that this is not an adequate reflection of the underlying project risks and does not account for the opportunity costs (Posner & Weisbach 2010). By extension, some private sector actors argue that a bond rate might be a more reasonable discount rate for local governments. New York City's *Climate Resiliency Design Guidelines* offers yet another alternative perspective on discount rate through their utilization of a present value coefficient that is a function of the project's useful life and the general project discount rate (2018, p. 33). In order to stimulate experimentation, the New York City guide also suggests that projects under $50 million should consider utilizing a qualitative CBA in order to allow for the determination of a variety of benefits that might defy quantification. In the future, it is anticipated that CBA/CEA will need to run a sensitivity analysis across a range of discount rates to find a rate that accommodates the values and outcomes implicit in public investment, as well as those market determinations of risk inherent in the cost of capital.

Evaluating funding sources

Assuming that a project has passed an initial screening among alternative project options, the next step is to evaluate a range of options for accessing and utilizing different funding sources. As previously referenced, providing some iterative sensitivity in the initial screening stage of project planning for funding and financing options is useful, if only to determine feasibility and the approximate cost of capital. However, local governments have a limited institutional capacity to seek and apply for funding and therefore they must be strategic in prioritizing their limited resources.

As referenced in Table 4.1, the City and County of San Francisco have developed a useful matrix for organizing and weighing relative options. In this case, a Likert score of 1–5 is assigned to each corresponding option and criteria so as to form a "heat map." A weighted average is then tabulated to rank each option. In particular, the revenue generating capacity of a source is triple weighted to internally weight the method. The first criterion relates to the source of the funds. The question here relates to whether the source of the funds is controlled by the local government or whether it is otherwise controlled by state, federal or private actors who impose varying levels of control and oversight over the use and administration of those funds.

A related factor is the overall costs of those funds. As such, one must determine whether the funds can be leveraged through bonds or other securities; whether there are matching requirements; and, whether the taxable status of the funds is beneficial. In some cases, the source of the funds are grants and the cost is nearly free but for the administration costs. In this regard, the administrative complexity of certain funding sources is important given that some sources may require complex and costly studies or may impose burdensome appropriations or procurements processes. In other cases, even if the funds are nearly cost-free, the funds may not be particularly flexible and may come with conditions and stipulations that have collateral impacts on the ability to securitize or combine with other sources of funds.

Timing is another critical factor. Timing relates not only to an alignment of project phases and investment cycles, but also to the lead time associated with applying for or otherwise developing funding. For instance, sources that require legislative authority or voter consent may require many years of effort and expenses. The *Finance Guide for Resilient By Design Bay Area Challenge Design Teams* (RBD 2017) provides an authoritative perspective on the mechanics of mobilizing and aligning constituencies in California in order to develop various sources of funding. As noted in the guide and referenced in Table 4.1, an additional factor relates to the sustainability of funding sources. For instance, post-disaster grants are not reliable or sustainable, but certain voter-driven or state-sanctioned funding allocations may provide the long-term continuity necessary to bring down transactional costs.

Table 4.1 Example of funding strategies heat map

Rank	Funding strategy	Source of funds	Revenue-generating potential***	Cost of funds	Long-term sustainability	Flexibility of funds	Timing	Trade-offs for other city needs	State/Federal political feasibility	Local/Regional political feasibility	Administrative complexity	Equity/Cost burden	Weighted average
1	Local property tax increment from IFDs												4.77
2	Community Facilities District (CFD)												4.46
3	USACE – CAP 103 Program												4.38
4	State property tax increment from IFDs												4.31
5	General Obligation (G.O.) Bonds												4.23
6	Cap-and-trade program funding												4.23
7	State resilience G.O. Bond												4.15
8	Sales tax increase												4.08
9	Hotel assessment												4.00
10	Increased parking revenues												4.00
11	Assessment District (AD)												3.85
12	USACE – general investigation												3.77
13	Philanthropy												3.77

(Continued)

Table 4.1 (Continued)

Rank	Funding strategy	Source of funds	Revenue-generating potential***	Cost of funds	Long-term sustainability	Flexibility of funds	Timing	Trade-offs for other city needs	State/Federal political feasibility	Local/Regional political feasibility	Administrative complexity	Equity/Cost burden	Weighted average
14	Historic tax credits												3.62
15	Tax/Fee on marina use												3.54
16	Cruise tickets surcharge increase												3.46
17	Advertising												3.46
18	RM3 – bridge tolls program funding												3.38
19	Vehicle license fee (VLF) increase												3.38
20	Parcel tax												3.31
21	Naming rights												3.31
22	Congestion pricing												3.15
23	Public–private partnership (P3s)												3.08
24	Utility user tax Surcharge												2.92
25	Transit impact development fee												2.77
26	Transportation funding – TIFIA												2.77
27	Real estate transfer tax increase												2.69

#		Score
28	Surcharge on event tickets	2.62
29	Environmental impact bonds	2.62
30	Sale/Lease increment of port assets	2.62
31	Regional gas tax	2.46
32	Increased ferry charges	2.31
33	Hazard mitigation grants	2.31
34	Pension plan investment	2.31
35	Geologic Hazard Abatement Districts	2.23
36	Infrastructure trust bank	2.00
37	Transit pass transfer fee	1.00
38	Resilience Bonds/Insurance value capture	1.00

Source: City and County of San Francisco (2017).

Strength (5)
Partial strength (4)
Neither strength nor weakness (3)
Partial weakness (2)
Weakness (1)
*** Criteria triple weighted

An additional range of factors requires an evaluation of the political feasibility by and between federal and state and regional and local policies and
administrations. By some measure, this requires an assessment of the time
and costs of advocating for access to funding that otherwise represents some
measure of innovation. By another measure, this highlights the necessity
of evaluating trade-offs by and between not only competing policies but
also local priorities. In this sense, there is an opportunity cost whereas this
money could have been spent to address other more immediate challenges,
such as affordable housing and transportation accessibility. This opportunity cost may also represent a strategic path dependency in that it may also
reflect a reduction in the credit capacity to accommodate future necessary
adaptation investments.

The political feasibility and trade-off analysis highlights the final and per-
haps one of the most important factors – the social equity considerations and
cost burdens associated with any given funding source. Inequity may arise
by virtue of spatial or jurisdictional allocations of tax or assessment burdens,
but it may also arise incidental to institutionalized forms of historic inequity
related to segregation and environmental (in)justice. In some cases, such as
coastal tax increment financing, the inequity may manifest as a burden on
those that live and work on the coast even though inland populations may
also receive residual benefits. As previously referenced, one of the primary
concerns is the observed tendency of local governments to preference short-
term resilience investments framed in favor of property rights, economic
productivity and the maintenance of a tax base. While this is not an irrational
preference, the question is to what extent there are analytical references for
evaluating the distributional effects of such investments. Even in situations
where there is parity between beneficiaries and payees, the relative cost
burden may be an absolute burden on vulnerable populations who may not
have adequate resources to absorb such costs. Chapter 5 will provide some
additional guidance for the utilization of a variety of qualitative and quantitative methods for evaluating the distributional and equitable allocation of
costs and benefits.

Strategic positioning of projects

Once projects have been adequately screened and funding strategies have
been iteratively developed in coordination with project planning and development criteria, there is one final step in the analysis. This final step of the
analysis requires that the adaptation investment be contextualized within a
broader adaptation plan and strategy. Again, adaptation investments are generally not about the development of free standing projects derived to address
a singular climate impact. Rather, they are investments made in every day

Table 4.2 Example framework for a multi-criteria assessment of funding options

Risk	Treatment option	Economic efficiency										Consequences of actions/inaction					
		Effectiveness	Cost	Funding options	Time to implement	Duration	Technical feasibility	Human capacity	Regulatory impact	Community acceptance	Benefit	Climate change impact	Social impact	Environmental impact	Co-benefits	Secondary risks	Residual risk
1	A																
	B																
2	A								EXAMPLE								
	B																
	C																
	D																
3	A																
	B																

Source: Adapted from GSA (2015).

projects that require marginal investments in elements that increase the resilience performance and adaptive capacity of a particular asset. To this end, it is critical to reflect on the extent to which these disparate projects and project elements relate to one another in a broader strategic plan.

For instance, redundancy between the performance of different projects may operate to either reinforce network resilience or serve as a potential for economic waste. To this end, this strategic positioning may serve as a calibration of project elements. It is also an opportunity to document and update the assumptions of each of the projects as a source of intelligence for maintaining and updating a local adaptation plan. At some juncture in the future, as climatic conditions change, it will be necessary to revisit the assumptions and rationale for certain investments. This institutional learning is critical for developing precise indicators and measurements necessary for the underwriting of future projects.

5 Social equity considerations

As referenced throughout this guide, there are a variety of social equity implications to climate adaptation interventions, strategies and investments. These implications are often inherent to the concepts of resilience and adaptation themselves. For instance, the status quo biasing of resilience may operate to help advance the cohesiveness and viability of communities, but it may also perpetuate existing conditions in a manner that perpetuates "poverty traps" and otherwise operate to reinforce existing inequalities (Keenan 2016b). The challenge is to develop methodologies that can provide some measure of transparency for informing decision-making and public deliberations as to what and who should be the beneficiaries of adaptation investments (Tol, Downing, Kuik, & Smith 2004).

Quantitative weighting

Equity weights are adjustment factors applied to CBA estimates that adjust calculations of benefits relative to their underlying utility to populations with varying degrees of well-being – often measured by income. As a general proposition, equity weights

> reflect that a dollar to a [vulnerable household] is not the same as a dollar to a [non-vulnerable household]. That is, one cannot add up monetized welfare losses across disparate incomes. Instead, one should add up welfare losses and then monetize.
>
> (Anthoff, Hepburn & Toll 2009, p. 14)

In this case, contextual climate vulnerability may be defined as a wide range of social and environmental characteristics that shape the intersection of exposure, sensitivity and adaptive capacity (O'Brien, Eriksen, Nygaard, & Schjolden 2007). Vulnerability indicators often make up composite indices defined by race, gender, health and wealth attributes, as well as a variety of other attributes such as education attainment, access to mass transit, social cohesiveness, household size, civic participation, language, housing tenure, immigration status and sectorial employment. Given the potential unequal nature of losses for vulnerable populations, then, by extension, the "diminishing marginal utility of money implies that risk aversion and income differences should be taken into account [when] calculating social welfare benefits" (Kind, Botzen, & Aerts 2017, p. 1).

Although equity weights are generally not allowed to be utilized in CBAs in the U.S., it is worth highlighting this modified method for purposes of internal evaluation. Their utilization is not universally regarded by social welfare economists and public policy analysts, with some arguing that wealth distribution should be explicit in matters of legislative intent; equity (or distributional) weights are inefficient relative to alternatives such as direct transfers and tax policy; and, that equity weights undermine the universal value of the method. Although this later point is somewhat undermined by the subjective and diverse nature of preferences latent in CBA. In practice, it is argued that diminishing marginal utilities are accommodated through a discount rate. As previously discussed, there are many competing interests in the selection of discount rates that may undermine this utility argument. A second limitation is that equity weights are often highly sensitive to the "resolution of impact estimates" and may vary significantly by region and sample population (Anthoff, Hepburn, & Tol 2009, p. 836). Therefore, the method might be difficult to apply across large heterogeneous jurisdictions for large infrastructure projects with widely distributed benefits.

The determination of the adjustment factor within the CBA for each population cohort is subject to a variety of interpretations and logics that seek to rank "well-being" attributes with the underlying utility of a benefit (e.g., $1) distributed across a cohort (Adler 2016). A variety of social welfare functions (SWF) can be utilized to determine marginal elasticity, including utilitarian and isoelastic/Atkinson functions (id.). In conventional CBA, the marginal utility of benefits is assumed to be equal. In other words, the value of the benefit, as well as the costs of any losses, are assumed to be equal for everyone.

Utilizing a utilitarian framework found in CBA, the first step in calculating equity weights is to pick a value representing the elasticity of the marginal utility. This can be thought of as a value that represents the change in proportionate benefit relative to a population cohort's (e.g., income strata) well-being. Research suggests that these values should or could range from 0.5 to 1.5 (Squire & van der Tak 1992; Fankhauser, Tol, & Pearce 1997; Pearce, Atkinson, & Mourato 2006; H.M. Treasury 2014).

The next step is to utilize the elastic utility and equity weight functions represented in Figure 5.1. In the example illustrated in Table 5.1,

a.

$$U\,(Y) = \frac{Y^{1-\gamma}}{1-\gamma}$$

Elastic Utility Function

b.

$$\omega_{Yi} = (Y_i/Y_{avg})^{-\gamma}$$

Equity Weight Function

Figure 5.1 Functions for elastic utility and equity weights
Source: Adapted from Kind, Botzen, and Aerts (2017).

Table 5.1 Example of the incremental utility of $1 in determining equity weights

Income and utility	Average income	High income	Low income	Extremely low income
Y0	$100	$1,000	$50	$20
Y1	$101	$1,001	$51	$21
δY	$1	$1	$1	$1
U(Y0)	−0.2000	−0.0632	−0.2828	−0.4472
U(Y1)	−0.1990	−0.0632	−0.2801	−0.4364
U(δY)	0.0010	0.0000	0.0028	0.0108
Equity Weight (ωYi)	1.00	0.03	2.81	10.86

Note: Elasticity of marginal utility (γ) = 1.5.

it is assumed that elasticity value (γ) is 1.5. U is the utility and Y is the income. It is assumed that there are four well-being cohorts ranging from high income to extremely low income, including an average income cohort. The incremental utility ($U(\partial Y)$) equals 0.0010 for the average income cohort and 0.0028 and 0.0108 for the low and extremely low-income cohorts, respectively. Because the upper income is extraordinarily wealthy by comparison, the incremental utility is 0.00003. Equity weights (ω_{Yi}) are "normalized as the weights attached to the increase in income for different groups relative to that of the average," as reflected in Figure 5.1(b) (Kind, Botzen, & Aerts 2017, p. 8). Therefore, the result is a weight of 0.03 (0.00003/0.0010) for the high-income cohort and 10.86 (0.0108/.0010) for the very low-income cohort. The implication is that for every dollar ($1) of benefit to a high-income neighborhood, for instance, the benefit is valued at $0.03. That same dollar in a very low-income neighborhood is valued at $10.86.

Box 5.1 Applying equity weights to a flood levee CBA

In carrying forward the example highlighted in Table 5.1, it can be assumed that a flood levee project is being evaluated for a coastal section of a small California community. Because of financial constraints, the local government decided that it must phase the construction of a flood levee. The local government must determine which community will benefit from the first phase of investment. Levee A would protect a gentrifying low-income community and Levee B would protect a historically marginalized and highly vulnerable, very low-income community. Figure 5.2 highlights an application of the equity weights resulting from the calculations in Table 5.1. However, the weighted benefits do not reflect actual benefits that may accrue in some proportion to the maintenance of a tax base or other elements of economic output that are localized in either geography.

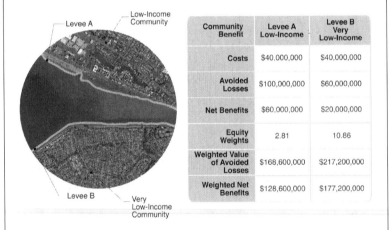

Community Benefit	Levee A Low-Income	Levee B Very Low-Income
Costs	$40,000,000	$40,000,000
Avoided Losses	$100,000,000	$60,000,000
Net Benefits	$60,000,000	$20,000,000
Equity Weights	2.81	10.86
Weighted Value of Avoided Losses	$168,600,000	$217,200,000
Weighted Net Benefits	$128,600,000	$177,200,000

Figure 5.2 Example equity weights applied in a CBA

Qualitative evaluation

Beyond equity weighting, adaptation investments can be qualitatively evaluated for their relative social and environmental impact, which is inclusive of criteria that address vulnerable populations. In this regard, much knowledge can be derived from the world of "Social Impact Investment," which can be defined as "actively placing capital in enterprises that generate social or environmental goods, services, or ancillary benefits, such as creating good jobs, with expected financial returns ranging from highly concessionary to above market" (Brest & Born 2013, p. 24). To guide investments, both the *Impact Reporting and Investment Standards* (IRIS) (GIIN 2018) and *Global Impact Investment Rating System* (GIIRS) (B-Lab 2018) provide standardized criteria for evaluating the potential success of any given investment.

However, these standards can be difficult to apply in adaptation investments where social impact may not be a primary determinant of legislative or design intent. For instance, the intent may be to reduce environmental exposure, which may indirectly reduce vulnerability and result in positive social impact. There may be very little associated assessment of a range of social and/or environmental impacts beyond those required by environmental quality review regulations. As a consequence, there is little guidance available for integrating a variety of values and co-benefits associated with resilience and adaptation investments.

One collection of resources that is available is the *Capital Project Screen, Guide & Tool* (2018) promulgated by a multi-institutional initiative known as SPARCC (Strong, Prosperous and Resilience Communities Challenge). The underlying set of resources provides evaluation criteria and methodologies (e.g., weighted MCA) for a variety of asset classes ranging from housing to commercial facilities and from infrastructure to green space. In particular, the screening criteria referenced in Table 5.2 require consideration for broader considerations for racial equity, health and climate change.

Table 5.2 SPARCC capital screening evaluation criteria (2018)

	Racial equity	*Health*	*Climate*
Project measurably improves social determinants of health and would be expected to reduce racial disparities in health outcomes.			
Project addresses other environmental determinants of health and would be expected to reduce racial disparities in preventable illness.			
Project responds to specific health needs of the community.			
Project is designed to impact racial equity outcomes identified by a collaborative table or a community-informed plan.			
Community is engaged in the design of the project and/or project is consistent with an existing community-informed plan.			

(Continued)

Table 5.2 (Continued)

	Racial equity	Health	Climate
Community is incorporated into the ownership, governance and/or asset-building aspects of the project.			
Project team has identified potential negative unintended racial equity outcomes and has developed strategy for mitigation.			
Project features a resilient and/or sustainable design with attention to energy and water efficiency.			
Project increases active or public transport options for residents and/or adds key neighborhood features and amenities.			
Project strengthens community members' resilience against impacts of climate change, emergencies and natural disasters.			
Project is informed by analysis of relevant data during and after development process in order to leverage project's impact.			
Project demonstrates consistency with collaborative work plan and theory of change.			

Source: SPARCC (2018).

The evaluation criteria require an assessment of not only positive but also negative internalities and externalities associated with any given investment. Implicit in the evaluation is a weighting that rewards co-benefits and collateral investment in sustainability and resilience. The utilization of tools such as this ultimately requires a determination of the minimal threshold for a quasi-empirical tally of weighted assignments. While tools are useful for organizing a range of projects and criteria, very often the sheer range of factors shaping decision-making requires a local government's qualitative determination for what is a just and equitable outcome by and between investment alternatives. In this regard, the value of climate leadership in adjudicating these equities can never be fully valued or discounted.

Specific to the State of California, there are a variety of indicators, tools and check-lists associated with vulnerability assessment. Vulnerability assessments are critical to evaluating the extent to which there may be ineq-uitable impacts or burdens on certain populations. These resources vary in their utility and applicability across a diversity of geographies and impacts. To bring order to this myriad of resources, OPR's *Defining Vulnerable Communities in the Context of Climate Adaptation* (2018) is a thorough guide for sorting through and selecting the most appropriate assessment tool or methodology. The guide also provides some useful check-lists for a range of considerations relating to social equity.

6   Private sector

The private sector plays an important role in California's collective adaptation to climate change. Given the scale of climate change, no single sector can be expected to carry the burden of climate change impacts in isolation. Public investments in resilience and hazard mitigation in a particular geography that are intended to protect a tax base will be limited in their effectiveness if private enterprises are not prepared to adapt to climate change in their own right. Private enterprises will need to advance their own adaptive capacity for identifying climate change signals that impact their bottomline, supply chains and markets. Thereafter, they will need to be able to develop and resource strategies that not only manage the risks but also take advantage of the opportunities. Inherent in these broader economic disruptions are opportunities to capture and create new markets. However, for those risks that do manifest, it is necessary for private enterprises to have the requisite investments in organizational resilience that allow for the continuity of business operations with minimal disruptions on consumers and communities.

Risks and opportunities

Companies have always adapted – a failure to do so means going out of business. However, private sector adaptation has historically been dependent on an understanding of supply and demand relating to consumer preferences and market exchanges. Climate change complicates an assessment of market conditions because it defies rationalization, or the predictability associated with market cycles. Beyond bankruptcy, a failure to fully internalize the impacts of climate change on a business model could operate to impair assets, increase liabilities and decrease revenue. For publicly and privately held companies alike, these impacts may collectively operate to reduce shareholder and enterprise value.

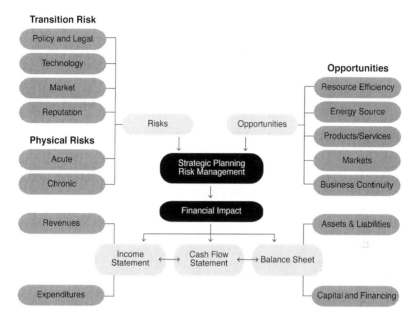

Figure 6.1 Climate-related risk, opportunities and financial impacts
Source: Adapted from FSB (2017a).

Understanding and classifying the range of risks and opportunities associated with climate change is a critical first step. As highlighted in Figure 6.1, opportunities may stem from underlying operations and resource efficiencies; cheaper and more reliable energy supplies; new products and services for either shifting or new markets; and, a decrease in vulnerability and exposure associated with organizational resilience investments in business continuity. As will be discussed, business continuity may also be a critical component for community resilience. Risks may be categorized into two broad categories: transition and physical risks (FSB 2017a). While this guide has primarily focused on physical risks for the public sector, transition risks are a relevant consideration for both public and private sectors.

The first component of transition risk relates to policy and legal considerations. As previously referenced, litigation stemming from disclosures may represent a potential underappreciated risk. In addition, changing policies outside of California may operate to undermine certain investment assumptions and market developments. Other transition risks relate to emergence of technologies that are designed to facilitate a broader economic transition to a low-to-no carbon economy. These new and emergent technologies may operate to accelerate the obsolescence of assets. By example, in the private sector, this may mean that power generation facilities based on heavy oil are no longer viable in terms of the levelized cost of electricity when compared with cheaper alternatives. For the public sector, this same comparatively obsolete facility may represent a challenge for obtaining additional debt capital that may be necessary for deferred maintenance or retrofitting. Recent research has suggested that if the Paris Accords were to be fully implemented, many sectors, including the energy sector, would be overvalued in what could otherwise be referenced as a "carbon bubble" (*The Economist* 2018).

The third and fourth components of transition risk relate to market demand and reputation. Again, these components may have relevance for both public and private sectors. For private enterprises, these risks may manifest not only in changing consumer demand based on emerging preferences, but it may also mean that it is necessary to accelerate research and development for new production, as well as increase capital allocations for the production and delivery of new goods and services. This can operate as both a risk and an opportunity. However, a failure to accommodate changing preferences and demands may manifest in a diminished reputation. These same considerations may also play out in the public sector where high-risk jurisdictions with poor quality infrastructure may be less desirable for corporate relocations or housing production. This desirability, or lack thereof, may be shaped by factors such as insurance coverage, local tax liabilities and infrastructure costs.

Whether it is an impact on income statements or balance sheets, the challenge for private enterprises is to develop an intelligence about the challenges and to adequately communicate those challenges. In some cases, that may mean formally communicating these risks and uncertainties through corporate disclosures (SEC 2010; Kahn 2017). In the absence of any public reporting and disclosure obligations, understanding and framing these risks is likely to be critical for obtaining adequate insurance coverage and for transferring risks through other contractual means. While minority shareholder groups are particularly vocal on these issues for public companies, it is likely only a matter of time before transition and physical climate risks become mainstream as part of underwriting and credit assessment processes.

Transition risks are often difficult to project or anticipate in the future, and, as such, private enterprises require ongoing intelligence about changing actors, operations, supply chains, value chains and markets to stay ahead of the curve. However, physical climate risks represent a level of empiricism that warrants short-, mid- and long-term forward-looking assessments. This analysis may include: (i) an assessment of the number of business lines and facilities exposed to specific hazards and/or stresses; (ii) projected changes in production, revenues, operational expenditures and capital expenditures due to considerations relating to exposure and vulnerability; (iii) probabilistic estimates of extreme events on operations, production, suppliers, customers or markets; and, (iv) a distillation of average annual losses (AAL) from climate change (Mazzacurati, Firth, & Venturini 2018).

There are two approaches for advancing these types of assessments for first and second-order impacts and effects. The first approach is to assess exposure or potential losses based on probabilities of the occurrence and depth of certain impacts. This could be as simple as determining the expected values (EV) of any given event on any given investment, asset or portfolio. However, a more sophisticated approach is to undertake a Value-at-Risk (VaR) analysis. VaR is a collection of various quantitative methodologies for applying probability to determine market and credit risk exposure of an institution. The output of this analysis is an estimate of the maximum loss that can occur with x% confidence over a holding period of t days (Choudhry 2013). As highlighted in Table 6.1, applications of probabilistic approaches in practice are limited to time horizons of no more than 20 years. However, for certain long-term investments, it might be appropriate to extend first-order estimates where probabilities have high degrees of certainty beyond 20 years but within the useful life of the investment or asset (e.g., sea level rise).

The other approach is scenario planning, which is loosely defined as an ordered process by which actors stress-test – often through complex narratives – what are assumed to be internally consistent perceptions and assumptions about possible futures and the extent to which those futures challenge and shape assumptions and emergent strategies (Lindgren & Bandhold 2003). Scenario planning may result in a quantitative range or a qualitative narrative about possible scenarios and the chain of events and consequences defining those scenarios. Scenario planning is useful for when there is not a reliable confidence in known probabilities or for when probabilities or distributions are simply not known or able to be constructed. For this approach to be most

Table 6.1 Timeframe and approach to assessing physical climate risks

	Recommended timeline	*Approach for first-order impacts*	*Approach for second-order impacts*
Short term	3–5 years	Probabilistic	Scenario analysis
Medium term	5–20 years	Probabilistic	Scenario analysis
Long term	20+ years	Scenario analysis	Scenario analysis

Source: Mazzacurati et al. (2018).

effective, it is necessary to follow a variety of known emissions and climate scenarios, as well as variants of these scenarios that represent extreme possible scenarios or otherwise account for degrees of uncertainty (Star et al. 2016). This may require incorporating data from a wide variety of sources both internal and external to an organization (Mazzacurati et al. 2018).

Table 6.2 provides a range of assumptions and parameters that may drive a starting point for a scenario analysis in a private enterprise. These assumptions speak to the selection of underlying drivers that are responding to or preparing for climate change impacts or transitions. In this regard, the interaction between climate mitigation and adaptation goals is conceptualized to

Table 6.2 Key considerations for conducting a scenario analysis

Parameters/Assumptions	*Analytical choices*	*Business impacts/effects*
Discount rate: What discount rate does the organization apply to discount future value?	**Scenarios:** What scenarios does the organization use for transition impact analysis and which sources are used to assess physical impact both for central/base case and for sensitivity analyses?	**Earnings:** What conclusions does the organization draw about impact on earnings and how does it express that impact?
Carbon price: What assumptions are made about how carbon price(s) would develop over time (within tax and/ or emissions trading frameworks), geographic scope of implementation, whether the carbon price would apply only at the margin or as a base cost, whether it is applied to specific economic sectors or across the whole economy and in what regions? Is a common carbon price used (at multiple points in time?) or differentiated prices? Assumptions about scope and modality of a CO_2 price via tax or trading scheme?	**Quantitative vs. qualitative or "directional":** Is the scenario exercise fully quantitative or a mix of quantitative and qualitative?	**Costs:** What conclusions does the organization draw about the implications for its operating/production costs and their development over time?

Parameters/Assumptions	Analytical choices	Business impacts/effects
Energy demand and mix: What would be the resulting total energy demand and energy mix across different sources of primary energy, e.g. coal/oil/gas/ nuclear/renewables (sub-categories)? How does this develop over time assuming supply/end-use efficiency improvements? What factors are used for energy conversion efficiencies of each source category and for end-use efficiency in each category over time?	**Timing:** How does the organization consider timing of implications under scenarios? E.g. is this considered at a decadal level (2020; 2030; 2040; 2050)?	**Revenues:** What conclusions does the organization draw about the implications for the revenues from its key commodities/products/ services, and their development over time?
Price of key commodities/ products: What conclusions does the organization draw, based on the input parameters/ assumptions about the development over time of market prices for key inputs, energy (e.g., coal, oil, gas, electricity)?	**Scope of application:** Is the whole analysis applied to the whole value chain (inputs, operations and markets) or just direct effects on specific business units/operations?	**Assets:** What are the implications for asset values of various scenarios?
Macro-economic variables: What GDP rate, employment rate and other economic variables are used?	**Climate models/data sets:** Which climate models and data sets support the assessment of climate-related risks?	**Capital allocation/ investments:** What are the implications for capex and other investments?
Demographic variables: What assumptions are made about population growth and/or migration?	**Physical risks:** When assessing physical risks, which specific risks have been included and their severity (e.g., temperature, precipitation, flooding, storm surge, sea level rise, hurricanes, water availability/ drought, landslides, wildfires or others)? To what extent has the organization assessed the physical impact to its portfolio (e.g., largest assets, most vulnerable assets) and to what extent have physical risks been incorporated in investment screening and future business strategy?	**Timing:** What conclusions does the organization draw about development of costs, revenues and earnings across time (e.g., 5/10/20 years)?

(*Continued*)

Table 6.2 (Continued)

Parameters/Assumptions	*Analytical choices*	*Business impacts/effects*
Efficiency: To what extent are positive aspects of efficiency gains/clean energy transition/physical changes incorporated into scenarios and business planning?	To what extent has the impact on prices and availability in the **whole value chain** been considered, including knock on effects from suppliers, shippers, infrastructure and access to customers?	**Responses:** What information does the organization provide in relation to potential impacts (e.g., intended changes to capital expenditure plans, changes to portfolio through acquisitions and divestments, retirement of assets, entry into new markets, development of new capabilities, etc.)?
Geographical tailoring of transition impacts: What assumptions does the organization make about potential differences in input parameters across regions, countries, asset locations and markets?		**Business interruption due to physical impacts:** What is the organization's conclusion about its potential business interruption/ productivity loss due to physical impacts, both direct effects on the organization's own assets and indirect effects of supply chain/ product delivery disruptions?
Technology: Does the organization make assumptions about the development of performance/cost and resulting levels of deployment over time of various key supply and demand-side technologies (e.g., solar, wind, energy storage, biofuels, nuclear unconventional gas, electric vehicles and efficiency technologies in other key sectors including industrial and infrastructure)?		

Parameters/Assumptions	*Analytical choices*	*Business impacts/effects*
Policy: What are assumptions about strength of different policy signals and their development over time (e.g., national headline carbon emissions targets; energy efficiency or technology standards and policies in key sectors; subsidies for fossil fuels; subsidies or support for renewable energy sources)?		
Climate sensitivity assumptions: What are assumptions of temperature increase relative to CO2 increase?		

Source: Adapted from FSB (2017a).

be a key parameter for both synergy and conflict. Thereafter, the scenario planning is faced with a variety of analytical choices relating to scale, scope and timing. Together, the parameters and analytical choices are able to shape a narrative about a range of possible impacts and effects that represent both risks and opportunities. While scenario analysis is best utilized for long-term planning and strategy development, it is also useful for stress-testing current business continuity plans for a variety of events and circumstances that may or may not be formally memorialized.

Community resilience and business continuity

The connection between business continuity and community resilience is increasingly recognized as important not simply for the convenience of a customer base but also for the preservation and maintenance of a customer base. Disruptions and market failures of private enterprises can have significant negative implications for a local government's tax base, as well as its labor market. This is particularly true in California's rural counties where natural resource extraction and agricultural economies are highly sensitive to changes in climate and weather. In these cases, the impacts may not always be driven by a decline in yields. For instance, the occurrence of extreme heat waves may have measurable negative impacts on labor productivity for those laborers working outside (Heal & Park 2016).

In addition, consistent with the example in Box 1.2 in Chapter 1, local communities are often highly dependent on certain enterprises for the provision of critical goods and services. Whether it is a regional hospital, a local lumber supplier in a small town or a grocery store in an underserved neighborhood, communities are dependent on private enterprises not only as a matter of convenience but as a matter of economic viability. While there are often hazard classifications for critical facilities, there is generally little appreciation for the identification of critical private sector enterprises outside of the national defense sector. To this end, local governments and private enterprise have an opportunity to work together to advance processes that mutually support business continuity and community resilience goals (Tracey et al. 2017).

In recent years, FEMA has developed the Voluntary Private Sector Preparedness Program, known as PS-Prep, that integrated standards from various bodies to provide a comprehensive process for private enterprises, including small businesses (FEMA 2017). Specific to California, a variety of sectors have developed business continuity plans that are geared towards advancing community resilience, including the healthcare (California Association of Health Care Facilities 2014) and maritime sectors (California Maritime Security Council 2010). The California Governor's Office of Emergency Services (Cal OES) also has a variety of resources available for informing public and private sector business continuity and organizational resilience. Other approaches include the development of "Mainstreet Resilience Plans" that seek to inventory key businesses and assets and develop a joint working plan for prioritizing recovery and the reopening of business that communities rely on (City of New Orleans 2016). Prioritization of critical assets may include pharmacies, grocery stores, gas stations, hardware stores, logistics centers and doctors' offices. Interventions may include everything from a prioritization of energy service recovery to identification requirements for accessing recovery zones. Together, these plans highlight the co-dependency between communities and private enterprises and the value of working together to promote organizational and community resilience. In this regard, it is difficult to separate the value of resilience and adaptation investments by and between the public, civic and private sectors.

7   Climate services

Nearly every day, there are newly discovered insights about the drivers, impacts and effects of climate change. The sheer amount of information perpetuates classical information asymmetries that drive pricing and valuation in markets. By application, if local governments cannot adequately manage climate information, then it increases the likelihood that they will make suboptimal adaptation investments. This might mean not enough or too much insurance coverage or simply not having enough leverage in negotiating a risk premium. With both public and private sectors increasingly demanding more sophisticated climate services, a wide range of climate services providers have emerged in both the public and private sectors. Many of the private sector providers have developed highly specialized and proprietary products and services that local governments have little to no experience in procuring. As such, this chapter seeks to provide some considerations for qualifying and selecting climate services providers. Like with all goods and services, the goal is to maximize the public's investment in high quality outcomes that serve the intended purpose.

Indicators and activities

The public and private sectors are united in the necessity to seek, process and utilize the most up-to-date information concerning climate change drivers, changes in climate and impacts on physical and biological systems. As highlighted in Table 7.1, the State of California has a formal range (n = 36) of indicators associated with each of these facets of climate change (OEHHA 2018). As defined by the California Environmental Protection Agency (Cal EPA), "[i]ndicators are scientifically-based measurements that track trends in various aspects of climate change. Many indicators reveal discernable evidence that climate change is occurring in California and is having significant, measurable impacts in the state" (id., p. 9).

Table 7.1 Indicators of climate change in California

Climate change drivers

- Greenhouse gas emissions
- Atmospheric greenhouse gas concentrations
- Atmospheric black carbon concentrations
- Acidification of coastal waters

Changes in climate
- Annual air temperature
- Winter chill
- Extreme heat events
- Cooling and heating degree days
- Precipitation
- Drought

Impacts of climate change on physical systems
- Snowmelt runoff
- Snow-water content
- Glacier change
- Lake water temperature
- Coastal ocean temperature
- Sea level rise
- Dissolved oxygen in coastal waters

Impacts of climate change on biological systems

On humans
- Vector-borne diseases
- Heat-related mortality and morbidity

On vegetation
- Forest tree mortality
- Wildfires
- Ponderosa pine forest retreat
- Vegetation distribution shifts
- Changes in forests and woodlands
- Subalpine forest density
- Fruit and nut maturation time

On wildlife
- Spring flight of Central Valley butterflies
- Migratory bird arrivals
- Bird wintering ranges
- Small mammal and avian range shifts
- Effects of ocean acidification on marine organisms
- Nudibranch range shifts
- Copepod populations
- Sacramento fall-run Chinook salmon abundance
- Cassin's auklet breeding success
- California sea lion pup demography

Source: OEHHA (2018)

There are two practical challenges associated with indicators that must be addressed. The first challenge relates to adequately investing in the resources necessary for building an ongoing intelligence concerning observations and measurements that are sourced from the public, private and civic sectors. Inherent in this challenge is the requirement to translate complex scientific observations and findings to a variety of audiences that may not appreciate the statistical or scientific conditions and qualifications that come along with these communications. Likewise, interrelated and interdependent indicators are often measured along different time horizons that further compound the uncertainty clouding decision-making.

Beyond climate communications, the second challenge speaks to the analysis necessary to connect broad indicators with precise direct and indirect effects on assets and adaptation investments. Indicators do not in and of themselves speak to the adaptive capacity of a human response for either mitigating or adapting to such impacts. In addition, there may be very little understanding of the extent to which scientific findings have bearing on complex designed systems, if at all. For instance, it has only been in recent years that atmospheric CO_2 concentrations have been understood to accelerate carbonation processes leading to corrosion and degradation in concrete infrastructure (Stewart, Wang, & Nguyen 2011). Prior to this research, attention was primarily on indicators associated with water and saltwater. As such, investments in indicator measurements have to be complemented by investments in research and communications.

Products and services

The complexity of all of these tasks makes its difficult, if not impossible, for any one organization to comprehensively manage. To fill this gap, the climate services sector has emerged across a variety of public and private organizations. Climate services may be defined as scientifically based information and products that enhance users' knowledge and understanding about the impacts of climate on their decisions and actions (AMS 2012). From a practical point of view, climate services providers utilize primary climate research, such as those findings associated with indicators, to provide trends, projections and scenarios that allow users to anticipate a range of future events, climates and impacts. As represented in Figure 7.1, these providers also operate to aggregate and often partially validate best practices and emerging knowledge concerning the applications from the underlying science.

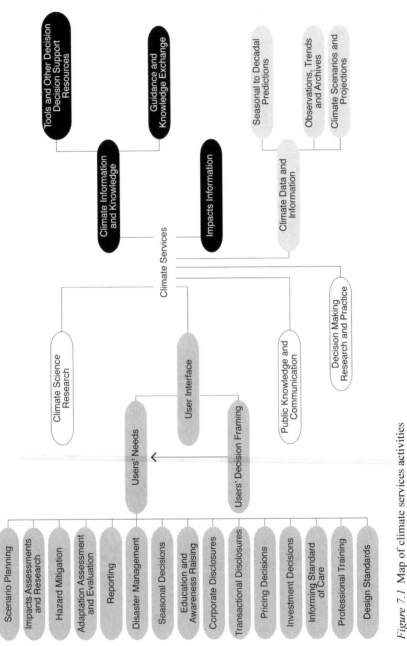

Figure 7.1 Map of climate services activities

Source: Adapted from UKCIP (2016).

This includes new and emerging models of climate communication. Climate communication strategies range from interface design for a range of consumers and decision-makers to the utilization of messaging that benefits from the latest advances in behavioral science (Galford et al. 2016). For local governments, some service providers focus on management protocols that seek to develop a capacity to manage streams of information across departments. By extension, they offer guidance on how to align interests and motivate compliance across various organizational units in order to inform emerging strategies for climate adaptation and resilience.

Much of the development of private sector climate services is based on the utilization of publicly derived data sets and packages. For instance, many private sector providers offer products that merely aggregate existing data sets that can be found for free in various deconstructed forms. Even early climate services firms, which offered agricultural insurance products and services based, in part, on proprietary hardware, originally took advantage of free data provided by the federal government. This is not necessarily a bad thing, but potential consumers should be sensitive to what tasks could be done in-house. For instance, some private sector firms have been known to offer sea level rise projections that are based on a simple extraction of data from NOAA's Sea Level Rise Viewer. This is a task that may be easily done in-house. By contrast, there are other firms that offer highly specialized services relating to relative sea level rise that are based on custom designed models that utilize advanced downscaled projections that benefit from independently derived geological and hydrological measurements and up-to-date digital elevation models. This example highlights why potential consumers should inquire about the underlying methodologies that are driving any analysis.

Increasingly, there are climate services providers that provide design, engineering and financial analysis that is inclusive of a variety of different climate indicators, models and considerations. This includes everything from specialized civil engineering for stormwater management design to structured finance experts that develop novel models to separate revenue and ownership elements to fully leverage public and private partnerships for hazard mitigation and adaptation projects. These downstream providers often work on well-established probability distributions and loss exceedance curves. This includes an accommodation of not only known parameters but also an accommodation of the uncertainty, deep uncertainty and general ignorance associated with future climates, impacts and effects. Currently, there are a number of cutting-edge providers that seek to construct probability distributions at a geographic and/or temporal resolution where previously downscaled models could not practically operate. The greater the resolution of the analysis, the greater the opportunity there is to address specific asset classes and sectors.

In recent years, there has been a proliferation of private sector providers that orient their services to a variety of specialized sectors and user demands. For instance, there are providers that offer seasonal forecasting to retailers who seek to optimize their purchasing of winter apparel. An unseasonably warm winter can have significant quarterly impacts on apparel retailers, which can also have an impact on local sales tax receipts. Closer to home, there are a variety of different providers that conduct impact assessments for a variety of different asset classes. Often these providers will specialize in either water or heat-related stresses. For example, several providers offer something called "flood scores" that are based on an indexed aggregation of various data points that produce an output measured in flood days per year. While research suggests that a specific number of floods days (e.g., $n = 26$) is the equivalent to effective inundation in terms of absolute economic loss, ultimately the decisions that are informed by these types of products and/or services are based solely on the judgment of the climate services consumer (Dahl, Spanger-Siegfried, Caldas, & Udvardy 2017). As a general proposition, nearly all climate services providers disclaim any liability from a consumer's reliance on their analysis. When it comes to predicting the future, that is still the purview of fortune-tellers.

Procurement considerations

Climate services providers rarely, if ever, take a position on an optimal path that a consumer should take. At best, providers will provide a range from which consumers may operate within. Like any other service providers, climate services providers are sensitive to warranting the precision or even the validity of their analysis. This means that consumers should be especially sensitive to labor hours committed to an analysis, as well as what data points that form the foundation of their analysis. If providers cannot sufficiently explain what they are doing in plain language, then this might be an indicator of future challenges for translating their work. When pushed for more details, sometimes providers will fall back on the proprietary nature of their modeling, which precludes them from giving away trade secrets. This should not be an absolute bar for providing clarification on the underlying methodologies and data points. To this end, it is often useful to ask a provider what they do differently from other providers that offer similar services. This is useful for understanding the experimental nature of their models and methods. The question for potential consumers is to what extent are they willing to serve as the test bed for untested products and services. In some cases, technological innovations may operate to discount products and services in favor of consumers.

For those providers with proprietary platforms and data sets, potential consumers should inquire about the terms and conditions that define the consumer's use of the data now and in the future. In some cases, the institutionalization of specific data sets or models may come with hefty licensing fees, even though the data was paid for by the consumer under a prior arrangement. As a general proposition, local governments should be mindful of the opportunity to learn from such engagements so that they may be able to emulate various data collection strategies and analytical methods in the future. In some cases, this may be impossible or beyond the competency of local staff. However, in many other cases, some data collection strategies may be cost-effective investments that may yield of variety of adaptation and climate mitigation co-benefits. A good example of this is quality assurance surveying among administrative units that have some measure of interdependency. For instance, there are a number of examples where energy and stormwater units found opportunities for collaborating on investing in elevating electric distribution that reduced transmission loss and increased the reliability of stormwater pumps and consumer electric service.

When procuring a climate services provider, potential consumers should give consideration to what are the fundamental problems and what information do they need and/or desire to have in order to make decisions that address those problems. Potential consumers should communicate these problems and information gaps when soliciting providers. Likewise, potential providers should be required to qualitatively address these such issues in a competitive bidding process or request for qualifications. In some cases, an internal inquiry might be just as effective as hiring an external third-party provider. In other cases, it might make more sense to hire a local university research lab to conduct applied research on a specific issue.

University labs offer a variety of advantages in terms of open access data and transparency. Likewise, these labs are often flexible and can easily modify the research as circumstances change and evolve. They may also be able to provide a screening analysis to identify the most appropriate specialized providers, as well as preexisting data sets and models that can help save consumers time and money in narrowing the scope of future arrangements. Local environmental planning, engineering, accounting and climate change consulting firms may also serve a similar role in narrowing an effective scope and developing a more effective procurement strategy. However, there is very little professional training and experience relating to climate change and this means that very large global firms tend to dominate the marketing landscape. This does not mean that these providers are necessarily the most qualified. There are a number of highly qualified small providers that are spread throughout California, the U.S. and the world.

Big or small, the challenge for potential consumers is to narrow the scope and to select the most qualified provider. Consumers should contextualize the acquisition of these products and services within a broader investment in the intelligence necessary to diagnose, prognose and resource climate change challenges. To this end, consumers have to think about the interoperability of these products and services within a range of internal and external constituencies. Consumers must also anticipate by some measure how they will learn from the process. This will reciprocally shape how they develop an adequate scope for procurement. The climate services sector will increasingly play an important role in informing local governments and the general public about everything from infrastructure investment to participatory planning – and maybe whether to buy a new winter coat this year. The challenge is to develop transparency and to maintain quality control in a manner that serves a variety of public interests.

8 Moving forward

This guide has provided a survey of a range of adaptation investment considerations that relate to assets, portfolios, funding sources, financing models and social equity tools. Moving forward, the challenge is to build the capacity of institutions and actors to develop the facility and intelligence necessary to underwrite and mobilize adaptation investments. A sensitivity to these issues, methods and models will be required across administrative units of local governments, as well as state agencies and private enterprises.

Asset managers will need to evaluate a range of risks and uncertainties across large portfolios of assets, and emergency managers will need to understand the ways and means of capital planning. Risk managers will need to move beyond historical data to anticipate probabilistic and nonprobabilistic stresses and shocks associated with climate change. Planners, designers and engineers will need to understand the asset management and risk management considerations defining LCCA and life-cycle performance. Bankers and bond investors will need to adjust the bottom line to account for lagging benefits and innovative investments in risk mitigation, resilience and adaptation. Community investors and advocates will need to continue to develop novel metrics of measuring and justifying a more just and equitable distribution of investment benefits. Ultimately, executive leadership and elected officials will need to understand that the economics of climate change can work for and against their respective constituencies. There are both costs and opportunities. To this end, this guide has provided a foundation for framing information that can support more effective, efficient and equitable decisions concerning the development of adaptation strategies, as well as the tactical deployment of capital among investment alternatives.

Climate change represents both a risk and an opportunity for California local governments and businesses. The broader discourse has historically focused on exposure and risk management. However, there are likely to be many opportunities for sustainable economic growth. For instance, new labor forces will be required to construct and maintain innovative infrastructure.

New materials and assemblies will need to be researched, designed, manu-factured and fabricated. With new products and services comes new markets both internal and external to California. In order to mobilize this economic engine and to take advantage of these climate change opportunities, new investments will need to be made in research, marketing and education. In this sense, climate adaptation investment is not simply about material investment and infrastructure. It also means investments in people and insti-tutions that must also adapt to a broad array of challenges ranging from an aging society to a shifting labor force. In order to accomplish this, society and institutions are tasked with developing a common conceptual and ana-lytical language that furthers more robust and transparent decision-making. This guide seeks to provide a foundation for this emerging momentum for primary education in climate adaptation tools, techniques, methodologies and processes. Only through a common understanding of the challenges can society adapt in manner that also captures the opportunities.

To accomplish a broader investment in education among public and pri-vate sector stakeholders, this guide is one of a number of resources that can be found on California's Adaptation Clearinghouse at resilientca.org. The California Adaptation Clearinghouse is intended to serve as a platform for providing ongoing intelligence about the latest advancements and innova-tions in not only adaptation investment, but also updates concerning impact assessment, vulnerability assessment, adaptation planning and develop-ment, execution challenges, pilot innovations and stories that share the suc-cesses and failures along the way.

Future adaptation investment research will be tasked with developing local reporting and audit controls; modeling financial conduits; and orga-nizing models for coordination that collectively speak to the scalar deploy-ment and management of adaptation investments. Research will need to be advanced across a variety of asset classes to evaluate everything from system performance of infrastructure to rates and degrees of material degradation. Corresponding research will be challenged to develop controls for report-ing and accounting as part of a broader intelligence about the emergence of unanticipated primary and secondary impacts. The operations of public services and the private delivery of goods and services will be subject to new understandings concerning supply chains and interdependencies from the shocks and stresses of climate change. Research will not only be tasked with understanding and assessing vulnerabilities, but it will be tasked with cataloging, classifying and communicating adaptation strategies and inter-ventions across a variety of sectors and actors. Finally, future research will need to advance processes, rules and institutions that help advance equitable determinations of the allocation of limited resources in the advancement of addressing the needs of disadvantaged and low-income communities.

The State of California is uniquely positioned to provide not only a vision but a pathway for a shared climate future that repositions risks as opportunities for a more sustainable environment and quality of life. Local governments, state agencies and local private enterprise are the front lines of broader adaptation to climate change. As a leader in addressing climate change, California has the opportunity to both shape and learn from a variety of jurisdictions and enterprises. This guide has provided a variety of methods and resources that mark the beginning of a long and potentially rewarding journey of climate change. Through a shared commitment to adaptation investment, there is endless opportunity to capture returns that will yield many benefits for future generations.

Bibliography

Adger, W. N., Arnell, N. W., & Thompkins, E. (2005). Successful Adaptation to Climate Change Across Scales. *Global Environmental Change*, 15, 77–86.

Adler, M. D. (2016). Benefit–Cost Analysis and Distributional Weights: An Overview. *Review of Environmental Economics and Policy*, 10(2), 264–285.

AghaKouchak, A., Ragno, E., Love, C., & Moftakhari, H. (2018). *Projected Changes in California's Precipitation Intensity-Duration-Frequency Curves*. California's Fourth Climate Change Assessment, California Energy Commission (CEC). Sacramento, CA: CNRA.

Amador, C. (2016). *Enhanced Infrastructure Finance Districts: Resource Guide to EIFDS*. California Community Economic Development Association. Retrieved from http://cceda.com/wp-content/uploads/EIFD-Resource-Guide-Feb-20161.pdf

Amekudzi, A., Crane, M., Springstead, D., Rose, D., & Batac, T. (2013). *Transit Climate Change Adaptation Assessment/Asset Management Pilot for the Metropolitan Atlanta Rapid Transit Authority*. FTA Report No. 0076. Washington, DC: Federal Transit Administration.

American Meteorological Society (AMS) (2012). *Climate Services: A Policy Statement of the American Meteorological Science*. Retrieved from www.ametsoc.org/ams/index.cfm/about-ams/ams-statements/archive-statements-of-the-ams/climate-services/

American Water Works Association (AWWA) (2018). *Climate Change Resource Community*. Retrieved from www.awwa.org/resources-tools/water-knowledge/climate-change.aspx

Anthoff, D., Hepburn, C., & Tol, R. S. (2009). Equity Weighting and the Marginal Damage Costs of Climate Change. *Ecological Economics*, 68(3), 836–849.

B-Lab (2018). *GIIRS Funds*. Retrieved from http://b-analytics.net/giirs-funds

Blackman, J., Maidenberg, M., & O'Regan, S. V. (2018, April 8). Mexico's Disaster Bonds Were Meant to Provide Quick Cash After Hurricanes and Quakes. But It Often Hasn't Worked Out That Way. *Los Angeles Times*. Retrieved from www.latimes.com/world/mexico-americas/la-na-mexico-catastrophe-bonds-20180405-htmlstory.html

BlackRock (2016, September). *Adapting Portfolios to Climate Change: Implications and Strategies for All Investors*. Global Insights. New York, NY: BlackRock.

Blue Forest Conservation (2017). *Fighting Fire With Finance: A Roadmap for Collective Action* Retrieved from https://static1.squarespace.com/static/556a1885e4b0bdc6f079 4659/t/59c1157f80bd5e1cd855010e/1505826201656/FRB_2017_Roadmap_Report.pdf

Brest, P., & Born, K. (2013). When can impact investing create real impact. *Stanford Social Innovation Review*, 11(4), 22–31.

Buurman, J., & Babovic, V. (2016). Adaptation Pathways and Real Options Analysis: An Approach to Deep Uncertainty in Climate Change Adaptation Policies. *Policy and Society*, 35(2), 137–150.

California Air Resources Board (CARB) (2017a). *Methods to Assess Co-benefits for California Climate Investments: Air Pollutant Emissions.* California Climate Investments, Greenhouse Gas Reduction Fund. Retrieved from www.arb.ca.gov/cc/capandtrade/auctionproceeds/carb_air_pollutant_emissions_nrw.pdf

California Air Resources Board (CARB) (2017b). *Methods to Assess Co-benefits for California Climate Investments: Vehicle Miles Traveled.* California Climate Investments, Greenhouse Gas Reduction Fund. Retrieved from www.arb.ca.gov/cc/capandtrade/auctionproceeds/carb_vehicle_miles_traveled.pdf

California Air Resources Board (CARB) (2018a). *Draft Co-Benefit Assessment Methodology: Adaptation.* California Climate Investments, Greenhouse Gas Reduction Fund. Retrieved from www.arb.ca.gov/cc/capandtrade/auctionproceeds/draft_adaptation_am.pdf

California Air Resources Board (CARB) (2018b). *Draft Co-Benefit Adaptation Assessment Methodology: Asthma/Respiratory Disaster Incidence.* California Climate Investments, Greenhouse Gas Reduction Fund. Retrieved from www.arb.ca.gov/cc/capandtrade/auctionproceeds/draft_asthma_am.pdf

California Association of Geological Hazard Abatement Districts (CAGHAD) (2008). *Application Management for Sea Level Rise: Geological Hazard Abatement District.* Retrieved from http://ghad.org/wp-content/uploads/2018/04/Application-Management-for-Sea-Level-Rise_11-2008.pdf

California Association of Geological Hazard Abatement Districts (CAGHAD) (2011). *Geological Hazard Abatement Districts.* Retrieved from http://ghad.org/wp-content/uploads/2018/04/GHAD-document-for-CA-Assoc.-of-GHADs-website.pdf

California Association of Health Facilities (2014). *Continuity of Operations Plan Template.* Disaster Preparedness Program. Retrieved from www.calhospitalprepare.org/continuity-planning

California Association of Local Economic Development (CALED) (2017). *Primer on California's New Tax Increment Financing Tools.* Retrieved from www.cacities.org/Resources-Documents/Policy-Advocacy-Section/Hot-Issues/New-Tax-Increment-Tools/CALED-TIF-Primer-3–17-FINAL.aspx

California Department of Transportation (Caltrans) (2012). *Climate Change.* No. DP-30. Retrieved from www.dot.ca.gov/hq/oppd/rescons/guidelines/DP-30_Climate-Change.pdf

California Energy Commission (CEC) (2018). *Funding Opportunities for the Electric Program Investment Charge (EPIC) Program.* Retrieved from www.energy.ca.gov/contracts/epic.html

California Maritime Security Council (2010). *Port Recovery and Business Continuity Planning Considerations.* Retrieved from http://aapa.files.cms-plus.com/PDFs/Recovery%20Planning%20Guide.pdf

California Natural Resources Agency (CNRA) (2009). *California Climate Adaptation Strategy: A Report to the Governor of California in Response to Executive Order S-13-2008.* Sacramento, CA: CNRA.

California Natural Resources Agency (CNRA) (2016). *Safeguarding California: Implementation Action Plans.* Sacramento, CA: CNRA.

California Natural Resources Agency (CNRA) (2018a). *Safeguarding California Plan: 2018 Update.* Sacramento, CA: CNRA.

California Natural Resources Agency (CNRA) (2018b). *California Climate Change Assessments.* Retrieved from http://climatechange.ca.gov/climate_action_team/reports/climate_assessments.html

Chicago Transit Authority (CTA) (2016). *RED Ahead: Phase One & Transit TIF.* Retrieved from www.transitchicago.com/rpm/ttif/

Chinowsky, P., Helman, J., Gulati, S., Neumann, J., & Martinich, J. (2017). Impacts of Climate Change on Operation of the US Rail Network. *Transport Policy.* doi:10.1016/j.tranpol.2017.05.007

Choudhry, M. (2013). *An Introduction to Value-at-Risk.* London, UK: John Wiley & Sons.

City and County of San Francisco, Seawall Finance Work Group (2017). *Fortifying San Francisco's Great Seawall: Strategies for Funding the Seawall Resilience Project.* San Francisco, CA: Office of Resilience and Capital Planning, City and County of San Francisco. Retrieved from www.onesanfrancisco.org/sites/default/files/inline-files/Seawall%20Finance%20Work%20Group%20Report%20Final%20version_0.pdf

City of Berkeley (2016). *Resilience Strategy: A Plan to Advance Preparedness and Equity in Berkeley, a Community for Inclusiveness and Innovation.* Retrieved from www.cityofberkeley.info/uploadedFiles/City_Manager/Resilient_Berkeley/Berkeley_Resilience_Strategy_LowRes.pdf

City of New Orleans (2016). *Main Street Resilience Plan.* Retrieved from www.nola.gov/nola/media/One-Stop-Shop/CPC/Main-St-Resilience-Plan-FINAL-8-16-16.pdf

Climate Bonds Initiative (CBI) (2017). *Climate Bond Standard.* Retrieved from www.climatebonds.net/files/files/Climate%20Bonds%20Standard%20v2_1%20-%20January_2017.pdf

Coburn, A., Copic, J., Crawford-Brown, D., Foley, A., Kelly, S., Neduv, E., Ralph, D., Saidi, F., & Yeo, J. Z. (2015). *Unhedged Risk: How Climate Change Sentiment Impacts Investment.* Cambridge Institute for Sustainability Leadership. Cambridge, UK: University of Cambridge.

Coffel, E., & Horton, R. (2015). Climate Change and the Impact of Extreme Temperatures on Aviation. *Weather, Climate, and Society, 7*(1), 94–102.

Dahl, K. A., Spanger-Siegfried, E., Caldas, A., & Udvardy, S. (2017). Effective Inundation of Continental United States Communities With 21st Century Sea Level Rise. *Elementa: Science of the Anthropocene, 5*(1), 37. doi:10.1525/elementa.234

Daniel, J. S. (2017). Infrastructure: Roadways in a Rut. *Nature Climate Change, 7*(10), 694–695.

Davidson, J., Jacobson, C., Lyth, A., Dedekorkut-Howes, A., Baldwin, C., Ellison, J., Holbrook, N., Howes, M., Serrao-Neumann, S., Singh-Peterson, L., & Smith,

T. (2016). Interrogating Resilience: Toward a Typology to Improve Its Operationalization. *Ecology and Society*, 21(2), Art 27. Retrieved from www.ecologyand society.org/vol21/iss2/art27/

de los Reyes Jr, G., Scholz, M., & Smith, N. C. (2017). Beyond the "Win-Win" Creating Shared Value Requires Ethical Frameworks. *California Management Review*, 59(2), 142–167.

Department of Water Resources, State of California (DWR) (2008, January). *Economic Analysis Guidebook*. Retrieved from www.water.ca.gov/LegacyFiles/economics/downloads/Guidebook_June_08/EconGuidebook.pdf

District of Columbia (D.C.) Department of Energy & Environment (2018). *Stormwater Retention Trading Program*. Retrieved from https://doee.dc.gov/src

Dittrich, R., Wreford, A., & Moran, D. (2016). A Survey of Decision-making Approaches for Climate Change Adaptation: Are Robust Methods the Way Forward? *Ecological Economics*, 122, 79–89.

Donovan, A. (2016, May 26). *Community Development Financial Institutions Finding Innovative Ways to Build Climate Resilience*. Community Development and Financial Institutions Fund, U.S. Department of the Treasury. Retrieved from www.cdfifund.gov/impact/Pages/BlogDetail.aspx?BlogID=16

The Economist (2018, May 24). *Markets May Be Underpricing Climate-Related Risk*. Retrieved from www.economist.com/finance-and-economics/2018/05/24/markets-may-be-underpricing-climate-related-risk?

Englander, J. (2015). Climate Change and Rising Sea Level: Implications for Historic Preservation. *Forum Journal: National Trust for Historic Preservation*, 29(4), 3–8.

Enterprise Community Partners (2015). *Ready to Respond: Strategies for Multi-Family Resilience*. Washington, DC: Enterprise Community Partners, Inc.

Enterprise Community Partners (2018). *Disaster Recovery and Rebuilding*. Retrieved from www.enterprisecommunity.org/solutions-and-innovation/disaster-recovery-and-rebuilding

Environmental Protection Agency (EPA) Office of Air and Radiation/Office of Atmospheric Programs/Climate Change Division (2013, September 9). *Glossary of Climate Change Terms*. Retrieved from www.epa.gov/climatechange

Ernst & Young (2017). *How Should We Account for Climate Change? A Step-by-Step Guide to Implementing the Financial Stability Task Force Recommendations for Disclosing Climate Change Risk*. Retrieved from www.ey.com/Publication/vwLUAssets/EY-accounting-for-climate-change/$FILE/EY-accounting-for-climate-change.pdf

Fankhauser, S., Tol, R. S., & Pearce, D. W. (1997). The Aggregation of Climate Change Damages: A Welfare Theoretic Approach. *Environmental and Resource Economics*, 10(3), 249–266.

Federal Emergency Management Agency (FEMA) (2013). *Local Mitigation Plan Review Guide*. Washington, DC: U.S. Department of Homeland Security.

Federal Emergency Management Agency (FEMA) (2017). *The Voluntary Private Sector Preparedness Program: PS-Prep*. Retrieved from www.fema.gov/voluntary-private-sector-preparedness-program-ps-preptm-small-business-preparedness

Financial Stability Board, Bank of International Settlements (FSB) (2017a). *Recommendations of the Task Force on Climate-related Financial Disclosures: Final Report*. Basel: Bank of International Settlements.

Financial Stability Board, Bank of International Settlements (FSB) (2017b). *Technical Supplement: The Use of Scenario Analysis in Disclosure of Climate-related Risks and Opportunities*. Basel: Bank of International Settlements.

Fleurbaey, M., & Zuber, S. (2013). Climate Policies Deserve a Negative Discount Rate. *Chicago Journal of International Law*, 13(2), 565–595. Retrieved from https://chicagounbound.uchicago.edu/cjil/vol13/iss2/1

Fuss, S. (2016). Climate Economics: Substantial Risk for Financial Assets. *Nature Climate Change*, 6(7), 659.

Galford, G. L., Nash, J., Betts, A. K., Carlson, S., Ford, S., Hoogenboom, A., & Underwood, K. L. (2016). Bridging the Climate Information Gap: A Framework for Engaging Knowledge Brokers and Decision Makers in State Climate Assessments. *Climatic Change*, 138(3–4), 383–395.

Global Adaptation & Resilience Investment Working Group (GARI) (2016). *Bridging the Adaptation Gap: Approaches to Measurement of Physical Climate Risk and Examples of Investment in Climate Adaptation and Resilience*. Retrieved from http://427mt.com/wp-content/uploads/2016/11/GARI-2016-Bridging-the-Adaptation-Gap.pdf

Global Impact Investors Network (GIIN) (2018). *IRIS Metrics*. Retrieved from https://iris.thegiin.org/metrics

Gollier, C. (2012). *Pricing the Planet's Future: The Economics of Discounting in an Uncertain World*. Princeton, NJ: Princeton University Press.

Government of South Australia (GSA) (2015). *Climate Change Adaptation Guidelines for Asset Management*. Adelaide: Department of Planning, Transport and Infrastructure, Government of South Australia.

Hallegatte, S. (2009). Strategies to Adapt to an Uncertain Climate Change. *Global Environmental Change*, 19(2), 240–247.

Hallegatte, S., Shah, A., Brown, C., Lempert, R., & Gill, S. (2012). *Investment Decision Making Under Deep Uncertainty: Application to Climate Change*. World Bank Policy Research Working Paper No. 6193. Washington, DC: World Bank.

Hanak, E., & Lund, J. R. (2012). Adapting California's Water Management to Climate Change. *Climatic Change*, 111(1), 17–44.

Heal, G. (2017). The Economics of the Climate. *Journal of Economic Literature*, 55(3), 1046–1063.

Heal, G., & Park, J. (2016). Reflections–Temperature Stress and the Direct Impact of Climate Change: A Review of an Emerging Literature. *Review of Environmental Economics and Policy*, 10(2), 347–362.

Herman, J., Fefer, M., Dogan, M., Jenkins, M., Medellín-Azuara, J., & Lund., J. (2018). *Advancing Hydro-Economic Optimization to Identify Vulnerabilities and Adaptation Opportunities in California's Water System*. California's Fourth Climate Change Assessment, California Natural Resources Agency. Sacramento, CA: CNRA.

H.M. Treasury (2014). *The Green Book: Appraisal and Evaluation in Central Government*. London, UK: TSO.

Hosseini, S., Barker, K., & Ramirez-Marquez, J. E. (2016). A Review of Definitions and Measures of System Resilience. *Reliability Engineering & System Safety*, 145, 47–61.

Howard, L. S. (2018, January 9). Parametric Insurance Can Help Close Global Gap: Clyde & Co. Report. *Insurance Journal*. Retrieved from www.insurancejournal.com/news/international/2018/01/09/476651.htm

HR&A (2016). *LA River Enhanced Infrastructure Financing District Revenue Yield Analysis*. Prepared for the City of Los Angeles Economic & Workforce Development Department. Retrieved from http://clkrep.lacity.org/onlinedocs/2014/14-1349_misc_11-30-2016.pdf

Hsiang, S., Kopp, R., Jina, A., Rising, J., Delgado, M., Mohan, S., & Larsen, K. (2017). Estimating Economic Damage From Climate Change in the United States. *Science*, 356(6345), 1362–1369.

Huang, X. (2010). What Is Portfolio Analysis. In: *Portfolio Analysis: Studies in Fuzziness and Soft Computing*. Berlin: Springer.

International Capital Markets Association (ICMA) (2017). *Green Bond Principles*. Retrieved from www.icmagroup.org/assets/documents/Regulatory/Green-Bonds/GreenBondsBrochure-JUNE2017.pdf

[IPCC] Chambwera, M., Heal, G., Dubeux, C., Hallegatte, S., Leclerc, L., Markandya, A., McCarl, B. A., Mechler, R., & Neumann, J. E. (2014). Economics of Adaptation. In Field, C. B., Barros, V. R., Dokken, D. J., Mach, K. J., Mastrandrea, M. D., Bilir, T. E., Chatterjee, M., Ebi, K. L., Estrada, Y. O., Genova, R. C., Girma, B., Kissel, E. S., Levy, A. N., MacCracken, S., Mastrandrea, P. R., & White, L. L. (Eds.), *Climate Change 2014: Impacts, Adaptation, and Vulnerability. Part A: Global and Sectoral Aspects*. Contribution of Working Group II to the Fifth Assessment Report of the Intergovernmental Panel on Climate Change. Cambridge, UK: Cambridge University Press.

[IPCC] Field, C. B., Barros, V., Stocker, T. F., Qin, D., Dokken, D. K., Ebi, K., Mastrandrea, M. D., Mach, K. J., Plattner, G. K., Allen, S. K., Tignor, M., & Midgley, P. M. (Eds.) (2012). *Managing the Risks of Extreme Events and Disasters to Advance Climate Change Adaptation. A Special Report of Working Groups I and II of the Intergovernmental Panel on Climate Change*. Retrieved from www.ipcc.ch/pdf/special-reports/srex/SREX_Full_Report.pdf

[IPCC] Solomon, S., Qin, D., Manning, M., Chen, Z., Marquis, M., Averyt, K. B., Tignor, M., & Miller, H. L. (Eds.) (2007). *Climate Change 2007: The Physical Science Basis. Contribution of Working Group I to the Fourth Assessment Report of the Intergovernmental Panel on Climate Change*. Retrieved from www.ipcc.ch/pdf/assessment-report/ar4/wg1/ar4_wg1_full_report.pdf

Jorion, P. (2007). *Financial Risk Manager Handbook*. New York, NY: John Wiley & Sons.

Kahn, M. E. (2017). Requiring Companies to Disclose Climate Risks Helps Everyone. *Harvard Business Review*. Retrieved from https://hbr.org/2017/02/requiring-companies-to-disclose-climate-risks-helps-everyone

Keenan, J. M. (2015a). Sustainability to Adaptation and Back: A Case Study of Goldman Sachs' Corporate Real Estate Strategy. *Building Research & Information*, 43(6), 407–422. doi:10.1080/09613218.2016.1085260

Keenan, J. M. (2015b). Adaptive Capacity of Commercial Real Estate Firms to Urban Flooding New York City. *Journal of Water and Climate Change*, 6(3), 486–500. doi:10.2166/WCC.2015.097

Keenan, J. M. (2016a). Private Mainstreaming: Using Contract and Strategic Processes to Promote Organizational and Institutional Adaptation. *Projections: MIT Journal of Planning*, 12(1), 113–133.

Keenan, J. M. (2016b). The Resilience Problem: Part 1. In Graham, J., Blanchfield, C., Anderson, A., Carver, J., & Moore, J. (Eds.), *Climates: Architecture and the Planetary Imaginary*. New York, NY/Zurich: Columbia University Press/Lars Muller Publishers.

Keenan, J. M. (2018). Regional Resilience Trust Funds: An Exploratory Analysis for Leveraging Insurance Surcharges. *Environment Systems and Decisions*, 38(1), 118–139. doi:10.1007/s10669-017-9656-3

Keenan, J. M., Hill, T., & Gumber, A. (2018). Climate Gentrification: From Theory to Empiricism in Miami-Dade County, Florida. *Environmental Research Letters*, 13(5), 054001. doi:10.1088/1748-9326/aabb32

Keenan, J. M., King, D. A., & Willis, D. (2015). Understanding Conceptual Climate Change Meanings and Preferences of Multi-Actor Professional Leadership in New York. *Journal of Environmental Policy and Planning*, 18(3), 261–285.

Kind, J., Botzen, W. J., & Aerts, J. C. (2017). Accounting for risk aversion, income distribution and social welfare in cost-benefit analysis for flood risk management. *Wiley Interdisciplinary Reviews: Climate Change*, 8(2), e446.

Kingsborough, A., Jenkins, K., & Hall, J. W. (2017). Development and Appraisal of Long-term Adaptation Pathways for Managing Heat-risk in London. *Climate Risk Management*, 16, 73–92.

Kruitwagen, L., Madani, K., Caldecott, B., & Workman, M. H. (2017). Game Theory and Corporate Governance: Conditions for Effective Stewardship of Companies Exposed to Climate Change Risks. *Journal of Sustainable Finance & Investment*, 7(1), 14–36.

Kurth, M., Keenan, J. M., Sasani, M., & Linkov, I. (2018). Defining Resilience for the Building Industry for the U.S. *Building Research & Information*. doi:10.1080/09613218.2018.1452489

Larkin, S., Fox-Lent, C., Eisenberg, D. A., Trump, B. D., Wallace, S., Chadderton, C., & Linkov, I. (2015). Benchmarking Agency and Organizational Practices in Resilience Decision Making. *Environment Systems and Decisions*, 35(2), 185–195.

Legislative Analyst's Office (LAO) (1996). *Understanding Proposition 218*. Retrieved from www.lao.ca.gov/1996/120196_prop_218/understanding_prop218_1296.html

Legislative Analyst's Office (LAO) (2014). *A Look at Voter-Approval Requirements for Local Taxes*. Retrieved from www.lao.ca.gov/reports/2014/finance/local-taxes/voter-approval-032014.aspx

Levin, H. M., & McEwan, P. J. (2000). *Cost-Effectiveness Analysis: Methods and Applications* (Vol. 4). Thousand Oaks, CA: Sage Publications.

Levy, D. (2018). *Financing Climate Resilience: Mobilizing Resources and Incentives to Protect Boston From Climate Risks*. Boston, MA: University of Massachusetts, Boston, Sustainable Solutions Lab.

Lindgren, M., & Bandhold, H. (2003). *Scenario Planning: The Link Between Future and Strategy*. New York, NY: Palgrave MacMillan.

Linnenluecke, M. K., Birt, J., & Griffiths, A. (2015). The Role of Accounting in Supporting Adaptation to Climate Change. *Accounting & Finance*, 55(3), 607–625.

Linnerooth-Bayer, J., & Hochrainer-Stigler, S. (2015). Financial Instruments for Disaster Risk Management and Climate Change Adaptation. *Climatic Change*, 133(1), 85–100.

London School of Economics, Economics of Green Cities Program (LSE) (2016). *Co-Benefits of Urban Climate Action: A Framework for Cities*. London, UK: London School of Economics. Retrieved from https://www.c40.org/researches/c40-lse-cobenefits

Los Angeles County Regional Park and Open Space District (LAC) (2018). *Safe, Clean Neighborhood Parks and Beaches Protection Measure of 2016*. Retrieved from http://rposd.lacounty.gov/2016-ballot-measure/

Mandell, M. E. (2017). *Business Improvement Districts: Public/Private Conflicts*. League of California Cities. Retrieved from www.cacities.org/Resources-Documents/Member-Engagement/Professional-Departments/City-Attorneys/Library/2017/2017-Annual-Conference-CA-Track/M-Mandell-Business-Improvement-Districts-CA-Track.aspx

Mazzacurati, E., Firth, J., & Venturini, S. (2018). *Advancing TCFD Guidance on Physical Climate Risks and Opportunities*. Report of the European Bank for Reconstruction and Development. London, UK: Four Twenty Seven and Acclimatise.

Mearns, L. O. (2009, September 9). *Methods of Downscaling Future Climate Information and Applications*. White Paper. National Center for Atmospheric Research. Retrieved from www.narccap.ucar.edu/users/user-meeting-09/talks/Downscaling_summary_for_NARCCAP_Users_Meet09.pdf

Meerow, S., Newell, J. P., & Stults, M. (2016). Defining Urban Resilience: A Review. *Landscape and Urban Planning*, 147, 38–49.

Mercer (2011). *Climate Change Scenarios: Implications for Strategic Asset Allocation*. Public Report. London, UK: International Finance Corporation/Carbon Trust.

Metropolitan Transportation Commission (MTC) (2017). *Toll Bridge Seismic Retrofit Program*. Retrieved from https://mtc.ca.gov/our-work/invest-protect/toll-funded-investments/toll-bridge-seismic-retrofit-program

Moody's Investor Services (2017, November 28). *Evaluating the Impact of Climate Change on U.S. State and Local Issuers*.

Moser, S. C., Ekstrom, J. A., Kim, J., & Heitsch, S. (2018). *Adaptation Finance Challenges: Characteristic Patterns Facing California Local Governments and Ways to Overcome Them*. California's Fourth Climate Change Assessment, California Natural Resources Agency (CNRA). Sacramento, CA: CNRA.

Moulton Niguel Water District (2017). *Long Range Financial Plan Report*. Retrieved from www.mnwd.com/app/uploads/2017/08/2017-Long-Range-Financial-Plan-Report.pdf

Municipal Securities Rulemaking Board (MSRB) (2018). *Glossary of Municipal Securities Terms*. Retrieved from www.msrb.org/Glossary/Definition/CONDUIT-FINANCING.aspx

National Association of Insurance Commissioners (NAIC) Catastrophe Insurance Working Group of the Property and Casualty Insurance Committee (2010, November). *Catastrophe Computer Modeling Handbook*. Retrieved from www.naic.org/documents/prod_serv_special_ccm_op.pdf

National Institute of Building Sciences (NIBS) (2015). *Developing Pre-Disaster Resilience Based on Public and Private Incentivization*. Retrieved from www.nibs.org/resource/resmgr/MMC/MMC_ResilienceIncentivesWP.pdf

Nelson, A. C. (2018). *System Development Charges for Water, Wastewater, and Stormwater Facilities*. Baca Raton, FL: CRC Press.

Neumann, J. E., Price, J., Chinowsky, P., Wright, L., Ludwig, L., Streeter, R., & Martinich, J. (2015). Climate Change Risks to US Infrastructure: Impacts on Roads, Bridges, Coastal Development, and Urban Drainage. *Climatic Change*, 131(1), 97–109.

Nordhaus, W. D. (2011). The Economics of Tail Events With an Application to Climate Change. *Review of Environmental Economics and Policy*, 5(2), 240–257.

O'Brien, K., Eriksen, S., Nygaard, L. P., & Schjolden, A. (2007). Why Different Interpretations of Vulnerability Matter in Climate Change Discourses. *Climate Policy*, 7(1), 73–88. doi:10.1080/14693062.2007.9685639

Office of Emergency Services, Governor of the State of California (OES) (2018). *2018 State Hazard Mitigation Plan*. Retrieved from www.caloes.ca.gov/for-individuals-families/hazard-mitigation-planning/state-hazard-mitigation-plan

Office of Environmental Health Hazard Assessment, California Environmental Protection Agency (OEHHA) (2018). *Indicators of Climate Change in California*. Retrieved from https://oehha.ca.gov/climate-change/document/indicators-climate-change-california

Office of Management and Budget, President of the United States (OMB) (2016). *OMB Circular A-94: Discount Rates for Cost-Effectiveness, Lease Purchase and Related Analyses* (Revised November 2016). Washington, DC: Office of Management and Budget. Retrieved from https://obamawhitehouse.archives.gov/omb/circulars_a094/a94_appx-c

Office of Planning and Research, Governor of the State of California (OPR) (2017). *Planning and Investing for a Resilient California: A Guide Book for State Agencies*. Sacramento, CA: OPR.

Office of Planning and Research, Governor of the State of California (OPR) (2018). *Defining Vulnerable Communities in the Context of Climate Adaptation*. Sacramento, CA: OPR.

Office of the Controller, State of California (2018). *Guidelines Relating to Gas Tax Expenditures for Cities and Counties*. Sacramento, CA: Office of the Controller, State of California. Retrieved from www.sco.ca.gov/Files-AUD/gas_tax_guidelines_jan2018.pdf

Olsen, J. R. (2015). *Adapting Infrastructure and Civil Engineering Practice to a Changing Climate*. Reston, VA: American Society of Civil Engineers (ASCE).

Pacific Gas and Electric (PG&E) (2016). *Climate Change Vulnerability Assessments and Resilience Strategies*. Retrieved from www.pgecurrents.com/wpcontent/uploads/2016/12/PGE_climate_resilience_report.pdf

Pacific Gas and Electric (PG&E) (2018). *Resilience Communities Grant Program*. Retrieved from www.pge.com/en_US/residential/in-your-community/local-environment/resilient-communities/resilient-communities-grant-program.page

Page, E. A. (2007). *Climate Change, Justice and Future Generations*. Northampton, MA: Edward Elgar Publishing.

Pearce, D., Atkinson, G., & Mourato, S. (2006). *Cost-Benefit Analysis and the Environment: Recent Developments*. Paris: OECD Publishing.

Posner, E., & Weisbach, D. (2010). *Climate Change Justice*. Princeton, NJ: Princeton University Press.

Ranger, N., Reeder, T., & Lowe, J. (2013). Addressing 'deep'uncertainty over long-term climate in major infrastructure projects: four innovations of the Thames Estuary 2100 Project. *EURO Journal on Decision Processes*, 1(3–4), 233–262.

RAND Corporation (2018). *Robust Decision Making*. Retrieved from www.rand.org/topics/robust-decision-making.html

re:focus partners (2015). *A Roadmap for Resilience: Investing in Resilience, Reinvesting in Community*. Retrieved from www.refocuspartners.com/wpcontent/uploads/pdf/RE.invest_Roadmap-For-Resilience.pdf

re:focus partners (2017). *A Guide for Public-Sector Resilience Bond Sponsorship*. Retrieved from www.refocuspartners.com/wp-content/uploads/pdf/RE.bound-Program-Report-September-2017.pdf

Resilient By Design: Bay Challenge Finance Advisory Team (RBD) (2017). *Finance Guide for Resilient By Design Bay Area Challenge Design Teams*. San Francisco, CA: NHA Advisors. Retrieved from www.resilientbayarea.org/finance-guide

RMS (2017). *Metro Cat Re Ltd. Catastrophe Bond (Series 2017–1)*. Cedent/Sponsor: First Mutual Transportation Assurance Co. Retrieved from www.artemis.bm/deal_directory/metrocat-re-ltd-series-2017-1/

S&P Global (2017, October 17). *Understanding Climate Change Risk and U.S. Municipal Ratings*.

S&P Global (2018, March 20). *How Our U.S. Local Government Criteria Weather Climate Risk*.

Sahebjamnia, N., Torabi, S. A., & Mansouri, S. A. (2015). Integrated Business Continuity and Disaster Recovery Planning: Towards Organizational Resilience. *European Journal of Operational Research*, 242(1), 261–273.

Schoennagel, T., Balch, J. K., Brenkert-Smith, H., Dennison, P. E., Harvey, B. J., Krawchuk, M. A., & Turner, M. G. (2017). Adapt to More Wildfire in Western North American Forests as Climate Changes. *Proceedings of the National Academy of Sciences*, 114(18), 4582–4590.

Securities and Exchange Commission (SEC) (2010, February 8). *Commission Guidance Regarding Disclosure Related to Climate Change*. Retrieved from www.sec.gov/rules/interp/2010/33-9106.pdf

SPARCC (Strong, Prosperous and Resilience Communities Challenge) (2018). *Capital Project Screen, Guide & Tool*. San Francisco, CA: Enterprise Community Partners, the Federal Reserve Bank of San Francisco, the Low Income Investment Fund and the National Resources Defense Council. Retrieved from www.sparcchub.org/wp-content/uploads/2018/03/SPARCC-Capital-Screen-Guide-Survey-Tool.pdf

Squire, L., & van der Tak, H. G. (1992). *Economic Analysis of Projects: A World Bank Research Publication*. Baltimore, MD: John Hopkins University Press.

Star, J., Rowland, E. L., Black, M. E., Enquist, C. A., Garfin, G., Hoffman, C. H., & Waple, A. M. (2016). Supporting Adaptation Decisions Through Scenario Planning: Enabling the Effective Use of Multiple Methods. *Climate Risk Management*, 13(1), 88–94.

State of California Office of the State Treasurer. (2017). *Growing the U.S. Green Bond Market: Volume 1: The Barriers and Challenges*. Retrieved from http://treasurer.ca.gov/greenbonds/publications/reports/1.pdf

State Water Resources Control Board, California (SWRCB) (2018, February 27). *State Water Board Drought Year Water Actions: Conservation Water Pricing*. Retrieved from www.waterboards.ca.gov/waterrights/water_issues/programs/drought/pricing/

Stechemesser, K., Bergmann, A., & Guenther, E. (2015). Organizational Climate Accounting: Financial Consequences of Climate Change Impacts and Climate Change Adaptation. In Schaltegger, S., Zvezdov, D., Alvarez Etxeberria, I., Csutora, M., & Günther, E. (Eds.), *Corporate Carbon and Climate Accounting*. Zurich: Springer.

Stewart, M. G., Wang, X., & Nguyen, M. N. (2011). Climate Change Impact and Risks of Concrete Infrastructure Deterioration. *Engineering Structures*, 33(4), 1326–1337.

Stults, M. (2017). Integrating Climate Change into Hazard Mitigation Planning: Opportunities and Examples in Practice. *Climate Risk Management*, 17(1), 21–34.

Sturm, M., Goldstein, M. A., Huntington, H., & Douglas, T. A. (2017). Using an Option Pricing Approach to Evaluate Strategic Decisions in a Rapidly Changing Climate: Black–Scholes and Climate Change. *Climatic Change*, 140(3–4), 437–449.

Swiss Re (2016). *Ursa Re Ltd. Catastrophe Bond (Series 2016–1)*. Cedent/Sponsor: California Earthquake Authority. Retrieved from www.artemis.bm/deal_directory/ursa-re-ltd-series-2016-1/

Tol, R. S. (2018). The Economic Impacts of Climate Change. *Review of Environmental Economics and Policy*, 12(1), 4–25.

Tol, R. S., Downing, T. E., Kuik, O. J., & Smith, J. B. (2004). Distributional Aspects of Climate Change Impacts. *Global Environmental Change*, 14(3), 259–272.

Tracey, S., O'Sullivan, T. L., Lane, D. E., Guy, E., & Courtemanche, J. (2017). Promoting Resilience Using an Asset-Based Approach to Business Continuity Planning. *SAGE Open*, 7(2). doi:10.1177/2158244017706712

Transportation Research Board (TRB) (2018). *Climate Change: Activities at the Transportation Research Board*. Retrieved from www.trb.org/main/specialtypage climatechange.aspx

UKCIP (2016). Overview of Climate Adaptation Programs in Europe: Climate Services. Oxford, UK.: UKCIP.

U.S. Department of Energy (2016). *Climate Change and the Electricity Sector: Guide for Climate Change Resilience Planning*. Washington, DC: U.S. Department of Energy, Office of Energy Policy and Systems Analysis.

U.S. Department of Homeland Security (2015). *Communications Sector-Specific Plan: An Annex to the NIPP 2013*. Washington, DC: U.S. Department of Homeland Security, Office of Infrastructure Protection Services.

U.S. Department of Housing and Urban Development (USHUD) (2015). *National Disaster Resilience Competition (NDRC): Benefit Cost Analysis Q&A*. Retrieved from www.hudexchange.info/course-content/ndrc-nofa-specific-webinar-q-a-benefit-cost-analysis/NDRC-Benefit-Cost-Analysis-Q-and-A-Slides-2015-09-17.pdf

U.S. Department of Housing and Urban Development (USHUD) (2018). *National Disaster Resilience*. Retrieved from www.hudexchange.info/programs/cdbg-dr/resilient-recovery/

U.S. Department of Transportation (USDOT) (2016). *Adapting to Climate Change*. Retrieved from www.transportation.gov/mission/sustainability/adapting-climate-change

U.S. General Services Administration (USGSA) (2017). *Life Cycle Costing*. Retrieved from www.gsa.gov/node/81412

U.S. General Services Administration (USGSA) (2018). *Sustainable Facilities Tool: Plan*. Washington, DC: USGSA.

U.S. Global Change Research Program (USGCRP) (2018a). *Climate Change: Glossary*. Retrieved from www.globalchange.gov/climate-change/glossary

U.S. Global Change Research Program (USGCRP) (2018b). *What We Do: Assess the U.S. Climate*. Retrieved from www.globalchange.gov/what-we-do/assessment

U.S. National Institute of Standards and Technology (NIST) (1995). *Life-Cycle Costing Manual for the Federal Energy Management Program*. Gaithersburg, MD: U.S. Department of Commerce. Retrieved from www.fhwa.dot.gov/asset/lcca/010621.pdf

U.S. National Institutes of Standards and Technology (NIST) (2015a). *Community Resilience Planning Guide for Buildings and Infrastructure Systems* (Vol. 1). Washington, DC: U.S. Department of Commerce. doi:10.6028/NIST.SP.1190v1

U.S. National Institutes of Standards and Technology (NIST) (2015b). *Community Resilience Planning Guide for Buildings and Infrastructure Systems* (Vol. 2). Washington, DC: U.S. Department of Commerce. doi:10.6028/NIST.SP.1190v2

U.S. National Institute of Standards and Technology (NIST) (2017). *Investment Analysis Methods*. NIST Advanced Manufacturing Series 2005–5. Gaithersburg, MD.: U.S. Department of Commerce. Retrieved from https://nvlpubs.nist.gov/nistpubs/ams/NIST.AMS.200-5.pdf

United Nations International Strategy for Disaster Risk Reduction (UNISDR) (2017). *Terminology on Disaster Risk Reduction*. Retrieved from www.unisdr.org/we/inform/terminology

van der Pol, T. D., van Ierland, E. C., & Gabbert, S. (2017). Economic Analysis of Adaptive Strategies for Flood Risk Management Under Climate Change. *Mitigation and Adaptation Strategies for Global Change*, 22(2), 267–285.

Watkiss, P., Hunt, A., Blyth, W., & Dyszynski, J. (2015). The Use of New Economic Decision Support Tools for Adaptation Assessment: A Review of Methods and Applications, Towards Guidance on Applicability. *Climatic Change*, 132(3), 401–416.

Webster, M. (2008). Incorporating Path Dependency Into Decision-analytic Methods: An Application to Global Climate-change Policy. *Decision Analysis*, 5(2), 60–75.

White, L., & Grantham, J. (2017, August 25). Wall Street's Best Minds: Climate Change Offers Upside for Investors. *Barron's*. Retrieved from www.barrons.com/articles/grantham-climate-change-offers-upside-for-investors-1503686667

World Resources Institute (WRI) (2014, July). *Building Climate Equity: Creating a New Approach From the Ground Up*. Retrieved from www.wri.org/sites/default/files/building-climate-equity-072014.pdf

Appendix

Table A.1 Adaptation and resilience co-benefits with sustainability

		Co-benefits				
City goals	Policy actions	Economic	Social	Environmental	Climate adaptation benefits	Potential indicators
Resources Improve energy reliability	Increase engineering resilience of energy infrastructure	Cost savings from climate-related damages Reduced energy losses Stability of energy supply for production Short-term job creation from upgrading infrastructure	Stable delivery of essential services	Improved quality of air and water Reduced land contamination	Adaptive capacity to extreme weather	Annual occurrences and total hours of electric and gas grid disruption
Improve potable water supply reliability	Increase sources of water supply less subject to drought	Reduced economic impacts of water variability Security of water cooling for power stations	Security for the vulnerable in cities impacted by drought Improved access to clean water Improved sanitation	Reduced aquifer depletion	Reduced water shortages	Annual volume of graywater, stormwater and reused water
Provide basic services (electricity, water, solid waste, etc.)	Invest in compact urban growth and infrastructure resilience	Increased economic efficiency	Quality of life	Reduced health impacts	Uninterrupted or minimally disrupted utility supplies during extreme weather	Number of utility connections for electricity, water and wastewater

Improve food security	Maintain and increase urban agriculture	Revenue generation potential and local job creation, particularly for low-income groups; Possible reduction in food price volatility	Increased food security	Maintained and increased biodiversity and green space; Improved air quality from decreased food transport	Increased food security; Decreased urban flooding	Number of days of food supply before and after hazard events
Health Protect vulnerable populations	Improve health planning for heat waves	Increased labor productivity and economic production through reduced heat stress; Reduced direct health costs	Reduced mortality and health impacts from heat		Heat adaptation, with reduced energy demands	Heat-related mortality and morbidity in elderly and vulnerable residents
Protect vulnerable populations	Reduce impacts of flooding on health	Reduced damage costs; Reduced direct health costs; Increased property values	Reduced mortality and health impacts from flooding directly, from water-borne diseases and from contamination of drinking water	Reduced toxic contamination	Urban flooding adaptation, with less future economic, resource and energy demand	Incidences of disease outbreaks tied to flood conditions
Improve public health services	Improve disease information and protection	Reduced direct health costs	Reduced mortality and health impacts from specific diseases		Reduced post-event disease	Incidences of heat-related and vector-borne diseases and injuries

(*Continued*)

Table A.1 (Continued)

City goals	Policy actions	Co-benefits			Climate adaptation benefits	Potential indicators
		Economic	Social	Environmental		
Improve public safety and security	Improve disaster planning and management	Reduced damage costs Reduced disruption of energy, transport, water and communications networks Increased economic resilience	Reduced mortality and health impacts from disasters		Reduced impacts from extreme weather events	Average response time, first responders
Improve public health	Increase urban green space	Increased labor productivity and economic production through reduced heat stress Increased property values from proximity to green spaces	Reduced health impacts from heat and flooding Increased physical and mental health Enhanced public amenity	Improved biodiversity and ecosystems Maintained and increased green space	Reduced urban heat island effect and reduced flooding impacts	Percentage of urban land area dedicated to green/open space
Improve public health and safety	Increase awareness of climate impacts and promote positive and informed behaviors	Reduced impacts on productivity	Reduced impacts on vulnerable groups	Reduced environmental impacts through associated awareness	Increased community resilience and adaptive capacity	Consumer and professional training and education programs

Facilitate active lifestyles	Protect and increase green space for sports and recreation, schools/universities and tourism	Increased labor productivity, economic production and school performance through reduced heat stress Increased property values from proximity to green spaces Longer-term productivity benefits from healthy, educated population	Reduced health impacts from heat and flooding Improved health from physical activity Improved student performance Improved student mental performance	Increased biodiversity and ecosystem services Maintained and increased green space	Reduced urban heat island effect and reduced flooding impacts	Number of adults undertaking regular physical activity
Facilitate active lifestyles	Increase cycling and walking networks	Reduce traffic congestion	Improved physical health, such as reduction of cardiovascular disease, some cancers, diabetes and obesity Reduced mortality and injuries from road-related accidents Improved access quality of life	Improved air quality	Promote community resilience and adaptive capacity	Modal split (percentage of trips walking or cycling)

(*Continued*)

Table A.1 (Continued)

	City goals	Policy actions	Co-benefits			Climate adaptation benefits	Potential indicators
			Economic	Social	Environmental		
Mobility	Maintain and improve service levels	Flood resistant transport infrastructure (e.g. overhead cabling, raised tracks)	Reduced damage costs Reduced travel disruptions leading to productivity gains	Reduced impact of future climate change events		Reduced flooding impacts	Percentage of transit trips/services within five minutes of scheduled time during extreme weather events (precipitation, wind, etc.); costs of restarting transit services following flooding
	Maintain and improve service levels	Heat resistant rail infrastructure (e.g., high temperature construction materials)	Reduced damage costs Reduced travel disruptions leading to productivity gains	Reduce impact of future climate change events		Reduced impact of heat buckling	Percentage of transit trips/services within five minutes of scheduled time during extreme heat events

Buildings	Maintain and improve building stocks	Promote passive and active cooling strategies for new buildings, existing building retrofits	Increased labor productivity and economic production through reduced heat stress Reduced direct health costs	Reduced mortality and health impacts from heat Improved student performance	Heat resilient buildings in higher average temperature and extreme heat events	Heat-related morbidity and mortality
	Maintain and improve building stocks	Promote design strategies for new buildings, existing building retrofits to mitigate flood risks	Cost savings from reduced flooding damages Reduced direct health costs Increased labor productivity and economic production through reduced flood disruption Increased property values	Reduced mortality and health impacts Improved student performance	Flood resilient buildings	Level of insured, non-insured losses from flood-related property damage
	Maintain and improve building stocks	Promote design strategies for new buildings, existing building retrofits to mitigate extreme storm risks and ongoing stresses from climate change	Reduced damage costs Reduced direct health costs	Reduced mortality and health impacts from storms	Resilience from storms, high winds, chronic mold, inundation, etc.	Level of insured, non-insured losses from storm-related property damage

(Continued)

Table A.1 (Continued)

	City goals	Policy actions	Co-benefits			Climate adaptation benefits	Potential indicators
			Economic	Social	Environmental		
Economy	Reduced energy burden	Increase building energy efficiency (e.g., insulation)	Cost savings to building owners and occupiers Increase in property values through efficiency, "green" branded buildings Local job creation (mainly short term) Increased productivity (commercial buildings)	Health improvements from improved air quality Increased thermal comfort	Improved air quality Ecosystem services (green roofs)	Cold resilience (extreme weather events) housing	Elderly wintertime mortality; number of households in fuel poverty (after fuel costs they would be left with a residual income below the official poverty line)
	Maintain and improve levels of economic growth	Improve resiliency of infrastructure	Reduced damage costs Reduced disruption to utilities and travel	Reduced mortality Reduced health impacts of flooding Reduced number of householders forced from homes	Reduced water pollution Effective/ uninterrupted water collection and security	Reduced climate-related impacts on transport, energy, water, communications networks and buildings	Annual instances and total hours of mass transit and urban service disruption
	Maintain and improve levels of critical infrastructure	Improve stormwater management	Reduced costs from flood-related damages	Reduced mortality Reduced health impacts of flooding Reduced number of householders, businesses forced from homes, places of work	Reduced water pollution Water collection and security	Reduced flooding impacts	Average and peak receiving water quality measures (e.g., bacteria, suspended solids)

Maintain and improve levels of critical infrastructure	Improve flood defenses	Reduced costs from flood-related damages	Reduced mortality Reduced health impacts of flooding Reduced number of householders, businesses forced from homes, places of work	Erosion control Enhanced biodiversity Enhanced green space	Reduced flooding impacts	Level of investment in engineered flood defense
Maintain and improve levels of critical infrastructure	Improve livability through "green and blue" infrastructure	Improve livability through "green and blue" infrastructure	Recreation	Enhanced biodiversity and green space	Reduced heat and flooding impacts	Percentage of urban land area dedicated to green and blue infrastructure
Bring forward new development areas for urban expansion	Improve land planning and development control	Reduce damage costs Higher property values	Social inclusion Protection of more vulnerable groups	Floodplain areas protected	Reduce development risks in flood plains/flood zones	Number of development approvals in flood-prone areas
Stimulate economic growth	Establish cleantech business clusters and incentives	Innovation firm productivity growth in technology sector			Improved organizational and community resilience through utilization of adaptation-related goods and services	New firm formation annually by sector

(Continued)

Table A.1 (Continued)

City goals	Policy actions	Co-benefits			Climate adaptation benefits	Potential indicators
		Economic	*Social*	*Environmental*		
Stimulate economic growth	Increase internet communications technology in adaptation systems	Reduced damage costs Reduced disruption to transport, energy, water and communications networks Reduced health costs	Reduced mortality and health impacts		More effective pre-, during, and post-event communications and response	Annual instances and total hours of mobile telephony service disruption

Source: Adapted from LSE (2016).

Table A.2 Adaptation strategies for asset management

Types of adaptation options	Examples
No-Regrets – options that are worthwhile, justified (cost-effective).	• Avoiding building in high-risk areas (e.g., floodplains for new development or when relocating). • Conducting more frequent site inspections of infrastructure assets during extreme weather events. • Moving equipment and/or production elements to areas of lower risk. For example, moving back-up generators to areas less prone to flooding. • Developing new or update existing and internal standards/codes/guidelines to better consider climate change in infrastructure design. • Avoiding measures that may make it more difficult to adapt to a changing climate (i.e., design decisions should not inadvertently increase climate vulnerability over time).
Low-Regrets – options with relatively low costs and large benefits.	• Restricting the type and extent of development in high-risk areas (e.g., floodplains). • Adjusting the rainfall capacity of drainage infrastructure to withstand more rainfall without failure/flooding. • Including infrastructure protection measures into design (e.g., sea walls to protect coastal infrastructure that cannot be located in less vulnerable areas). • Incorporating redundancy in design to allow continued operations despite the loss of some elements of the service or network. • Transferring the risk to third parties (e.g., insurance parties where the risk is insurable).
Win-Win – options that have the desired result of minimizing risk but also deliver social, environmental and economic benefits.	• Improving preparedness and contingency planning to treat risk (e.g., setting up early warning systems or signage in flood, bushfire and heat wave events). • Building community capacity of risks (e.g., education and awareness campaign around public services during heat wave events). • Selecting more resilient materials and construction methods to make designs more robust in the face of increasing climate-related risk (e.g., replacing timber sleepers with concrete sleepers). • Designing critical components of a system to cope with increased potential system failure due to extreme events.
Flexible adaptation options – staging or delaying the implementation of options, particularly if risks alter over various time periods (e.g., short, medium or long term).	• Progressively withdraw affected assets in coastal areas. • Time introduction of adaptation options to coincide with planned maintenance and/or upgrading. • Building in a manner that allows retrofitting at a later date when climate change impacts may occur (e.g., allow width of a road corridor to raise for flooding at later date). • Designing for future climatic conditions if the asset is expected to operate for the next 50 years. • Alternatively, decrease the expected asset life to ten years and only consider current climate conditions. In some cases, shorter design life may offer greater flexibility to help manage uncertainty.

Source: Adapted from GSA (2015).

Table A.3 Potential opportunities to embed climate change into the FEMA Hazard Mitigation Planning Crosswalk

	Existing requirement per the FEMA Crosswalk	*Potential climate adaptation considerations*
Element A: Planning process	A1: Does the plan document the planning process, including how it was prepared and who was involved in the process for each jurisdiction?	This is a key element for developing due process considerations for participatory planning exercises and processes. This is also an opportunity to develop ground-up vulnerability assessments through community actors and stakeholders.
	A2: Does the plan document an opportunity for neighboring communities, local and regional agencies involved in hazard mitigation activities, agencies that have the authority to regulate development as well as other interests to be involved in the planning process?	Beyond local planning, this is a key element for motivating regional cooperation in the assessment of vulnerabilities, as well as preparatory and responsive actions and investments.
	A3: Does the plan document how the public was involved in the planning process during the drafting stage?	This is a key element for developing due process considerations for participatory planning exercises and processes.
	A4: Does the plan describe the review and incorporation of existing plans, studies, reports and technical information?	This is an opportunity to integrate climate mitigation, resilience and adaptation plans.
	A5: Is there discussion of how the community(ies) will continue public participation in the plan maintenance process?	This is a key element for developing due process considerations for participatory planning exercises and processes. This is also an opportunity to develop ground-up vulnerability assessments through community actors and stakeholders.
	A6: Is there a description of the method and schedule for keeping the plan current (monitoring, evaluating and updating the mitigation plan within a five-year cycle?)	With all adaptation planning exercises, this an opportunity to formalize the processes by which best available science is incorporated into planning and design processes, as well as to incorporate mechanisms that support leaning and management.

	Existing requirement per the FEMA Crosswalk	*Potential climate adaptation considerations*
Element B: Hazard identification and risk assessment	B1: Does the plan include a description of the type, location and extent of natural hazards that can affect each jurisdiction?	This is an opportunity to incorporate an all-hazards approach that looks not just at primary but also secondary impacts from climate change that are understood as both shocks and stresses.
	B2: Does the plan include information on previous occurrences of hazard events and on the probability of future hazard events for each jurisdiction?	Probabilistic and non-probabilistic climate change observations and projections can be utilized to frame a risk profile of not only occurrence likelihood but also the range and depth of associated impacts.
	B3: Is there a description of each identified hazard's impact on the community as well as an overall summary of the community's vulnerability for each jurisdiction?	This is an opportunity to develop bottom-up vulnerability assessments that can be maintained on an ongoing basis.
	B4: Does the plan address NFIP-insured structures within the jurisdiction that have been repetitively damaged by floods?	Repetitive loss considerations are increasingly a major concern not just for property owners, but also for local governments who provide urban services.
Element: Mitigation strategy	C1: Does the plan document each jurisdiction's existing authorities, policies, programs and resources and its ability to expand on and improve these existing policies and programs?	This is an opportunity to synthesize a variety of hazard, climate and emergency management plans.
	C2: Does the plan address each jurisdiction's participation in the NFIP and continued compliance with NFIP requirements, as appropriate?	This is an opportunity to look at existing designations for hazard zones, as well as those zones that are likely to be included in the future.
	C3: Does the plan include goals to reduce/avoid long-term vulnerabilities to identified hazards?	This is an opportunity to incorporate resilience design guidelines and long-term goals for incorporation into building codes and land use plans.

(Continued)

	Existing requirement per the FEMA Crosswalk	Potential climate adaptation considerations
	C4: Does the plan identify and analyze a comprehensive range of specific mitigation actions and projects for each jurisdiction being considered to reduce the effects of hazards, with emphasis on new and existing buildings and infrastructure?	This is an opportunity to include analysis that covers a variety of modeled climate futures.
	C5: Does the plan contain an action plan that describes how the actions identified will be prioritized (including cost-benefit review), implemented and administered by each jurisdiction?	This is an opportunity to include plan, project and portfolio evaluation criteria in a pre- and post-disaster investment context.
	C6: Does the plan describe a process by which local governments will integrate the requirements of the mitigation plan into other planning mechanisms such as comprehensive or capital improvement plans, when appropriate?	This is an opportunity to merge capital, asset and resilience plans into a comprehensive framework.
Element D: Plan review, evaluation and implementation	D1: Was the plan revised to reflect change in development?	This is an opportunity to consider how climate and extreme weather observations are shaping ongoing planning now and in the future.
	D2: Was the plan revised to reflect progress in local mitigation efforts?	This is a key component to ensure institutional learning from experimentation and innovation in hazard mitigation and resilience investments.
	D3: Was the plan revised to reflect changes in priorities?	This is a key component for synthesizing system and portfolio considerations with those values being driven by public policy.

	Existing requirement per the FEMA Crosswalk	*Potential climate adaptation considerations*
Element E: Plan adoption	E1: Does the plan include documentation that the plan has been formally adopted by the governing body of the jurisdiction requesting approval?	This is an opportunity to formalize adaptation and resilience planning instruments.
	E2: For multi-jurisdictional plans, has each jurisdiction requesting approval of the plan documented formal plan adoption?	This is an opportunity to formalize regional partnerships.

Source: FEMA (2013); Stults (2017).

*applicable to plan updates only

List of online resources

	Agency	Funding or financing source	Program	URL
Federal resources	CDC	Public Health	Climate Ready States and Cities Initiative	www.cdc.gov/climateandhealth/climate_ready.htm
	DHS	Disaster Risk Reduction and Resilience	Regional Resilience Assessment Program	www.dhs.gov/regional-resiliency-assessment-program
	DOE	Housing and Community Development	Property Assessed Clean Energy Program	www.energy.gov/eere/slsc/property-assessed-clean-energy-programs
	EPA	Housing and Community Development	Smart Growth Grants	www.epa.gov/smartgrowth/epa-smart-growth-grants-and-other-funding
	EPA	Water Management	Clean Water State Revolving Loan Fund	www.epa.gov/cwsrf
	EPA	Water Management	Water Infrastructure and Resiliency Finance Center	www.epa.gov/waterfinancecenter
	EPA	Water Management	Various Grants	www.epa.gov/grants
	FEMA	Disaster Risk Reduction and Resilience	Hazard Mitigation Grants	www.fema.gov/hazard-mitigation-grant-program
	FEMA	Disaster Risk Reduction and Resilience	Pre-Disaster Mitigation Program	www.fema.gov/hazard-mitigation-grant-program
	FEMA	Disaster Risk Reduction and Resilience	Flood Mitigation Assistance Program	www.fema.gov/hazard-mitigation-grant-program
	FTA	Transportation	Various Grants	www.transit.dot.gov/grants

Organization	Program	URL
ILG	Resource Search Engine	Sustainability Funding Resources	www.ca-ilg.org/post/sustainability-funding-resources
NOAA	Resource Search Engine	U.S. Climate Resilience Toolkit Funding Opportunities	https://toolkit.climate.gov/content/funding-opportunities
NOAA	Natural Systems and Green Infrastructure	Coastal Resilience Grants	www.fisheries.noaa.gov/grant/noaa-coastal-resilience-grants
NOAA	Natural Systems and Green Infrastructure	Office of Coastal Management Grants	https://coast.noaa.gov/funding/
NPS/IRS	Housing and Community Development	Federal Historic Preservation Tax Incentives	www.nps.gov/tps/tax-incentives.htm
USACE	Water Management	Planning Studies–Planning Assistance to States	www.nae.usace.army.mil/Missions/Public-Services/Planning-Assistance-to-States/
USACE	Water Management	Planning Studies–Flood Plain Management Services	www.nae.usace.army.mil/Missions/Public-Services/Planning-Assistance-to-States/
USACE	Water Management	Continuing Authorities Program	www.nae.usace.army.mil/Missions/Public-Services/Continuing-Authorities-Program/
USBR	Water Management	WaterSMART Water and Energy Efficiency Grants	www.usbr.gov/watersmart/weeg/index.html
USDA	Natural Systems and Green Infrastructure	Agricultural Conservation Easement Program	www.nrcs.usda.gov/wps/portal/nrcs/main/national/programs/easements/acep/
USDA	Agriculture and Working Lands	Natural Resources Conservation Service	www.nrcs.usda.gov/wps/portal/nrcs/main/national/programs/financial/
USDA	Agriculture and Working Lands	Risk Management Agency Crop Insurance	www.drought.gov/drought/node/149
USDOT	Transportation	Build America Bureau	www.transportation.gov/buildamerica
USDOT	Transportation	Better Utilizing Investments to Leverage Development Grants	www.transportation.gov/BUILDgrants

(*Continued*)

(Continued)

	Agency	Funding or financing source	Program	URL
	USFS	Fire and Forest Management	Various Grants	www.fs.usda.gov/detail/prc/tools-techniques/funding/?cid=STELPRDB5200611
	USFWS	Natural Systems and Green Infrastructure	Various Grants	www.fws.gov/r5fedaid/grants.html
	USHHS	Resource Search Engine	Various Grants	www.grants.gov/
	USHUD	Disaster Risk Reduction and Resilience	Community Development Block Grant Disaster Recovery Program	www.hudexchange.info/programs/cdbg-dr/
	USHUD	Housing and Community Development	Community Development Block Grants	www.hud.gov/program_offices/comm_planning/communitydevelopment/programs
State Resources	CARB	Disaster Risk Reduction and Resilience	Greenhouse Gas Reduction Fund California Climate Investments	ww2.arb.ca.gov/our-work/programs/california-climate-investments
	CA Adaptation Clearinghouse	Resource Search Engine	State Funding Opportunities	http://resilientca.org/topics/investing-in-adaptation/#resources
	CA Senate	Natural Systems and Green Infrastructure	California Senate Bill 5: California Drought, Water, Parks, Climate, Coastal Protection, and Outdoor Access For All Act of 2018	https://leginfo.legislature.ca.gov/faces/billTextClient.xhtml?bill_id=201720180SB5
	CA State Parks	Housing and Community Development	Office of Grants and Local Services Program	www.parks.ca.gov/?page_id=1008
	CAL FIRE	Fire and Forest Management	Fire Prevention Grant Program	www.fire.ca.gov/grants/grants

Agency	Category	Program	URL
CAL FIRE	Fire and Forest Management	California Climate Investments Forest Health Grant Program	www.fire.ca.gov/grants/grants
CAL FIRE	Fire and Forest Management	California Forest Improvement Program	www.fire.ca.gov/grants/grants
CAL FIRE	Housing and Community Development	Urban and Community Forestry Grants	www.fire.ca.gov/grants/grants
Caltrans	Transportation	California Active Transportation Program	www.dot.ca.gov/hq/LocalPrograms/programInformation.htm
Caltrans	Transportation	Low Carbon Transit Operations Program	www.dot.ca.gov/drmt/splctop.html
Caltrans	Transportation	Sustainable Communities Grants	www.dot.ca.gov/hq/tpp/grants.html
Caltrans	Transportation	Transit and Intercity Rail Capital Program	www.dot.ca.gov/drmt/sptircp.html
Caltrans	Transportation	Transportation Planning Grants for Adaptation Planning	www.dot.ca.gov/hq/tpp/grants.html
CARB	Resource Search Engine	Funding Wizard	https://fundingwizard.arb.ca.gov/
CCC	Natural Systems and Green Infrastructure	Climate Ready Program	http://scc.ca.gov/climate-change/climate-ready-program/
CCC	Natural Systems and Green Infrastructure	Proposition 1 Grants	http://scc.ca.gov/2018/04/06/coastal-conservancy-prop-1-grant-solicitation/
CDFA	Agriculture and Working Lands	Healthy Soils Program	www.cdfa.ca.gov/grants/index.html
CDFA	Agriculture and Working Lands	State Water Efficiency and Enhancement Program	www.cdfa.ca.gov/grants/index.html
CDFA	Agriculture and Working Lands	Sustainable Agricultural Lands Conservation Program	www.conservation.ca.gov/dlrp/SALCP

(Continued)

(Continued)

Agency	Funding or financing source	Program	URL
CWC	Natural Systems and Green Infrastructure	Water Management	www.wildlife.ca.gov/Conservation/Watersheds/WSIP
DPR	Natural Systems and Green Infrastructure	Land and Water Conservation Fund Program	www.parks.ca.gov/?page_id=21361
DBW	Natural Systems and Green Infrastructure	Shoreline Erosion Control & Public Beach Restoration Grants	www.parks.ca.gov/?page_id=28766
CDFW	Natural Systems and Green Infrastructure	Environmental Enhancement Fund	www.wildlife.ca.gov/Grants
CDFW	Natural Systems and Green Infrastructure	State Wildlife Grants	www.wildlife.ca.gov/Grants
CDFW	Natural Systems and Green Infrastructure	Wetlands Restoration for Greenhouse Gas Reduction Grant Program	www.wildlife.ca.gov/Grants
CDFW	Natural Systems and Green Infrastructure	Fisheries Restoration Grants Program	www.wildlife.ca.gov/Grants
CDFW	Natural Systems and Green Infrastructure	Proposition 1 Watershed Restoration Grant Program	www.wildlife.ca.gov/Grants
CEC	Energy	Energy Conservation Assistance Act Low Interest Loans	www.energy.ca.gov/efficiency/financing/
CNRA	Housing and Community Development	Urban Greening Program	http://resources.ca.gov/grants/wp-content/uploads/2018/01/Urban-Greening-Guidelines-Round-Two.pdf
CNRA	Water Management	Urban Rivers Grant Program	http://resources.ca.gov/grants/california-urban-rivers/

CNRA	Transportation	Environmental Enhancement and Mitigation Program	http://resources.ca.gov/grants/
DPR	Natural Systems and Green Infrastructure	Habitat Conservation Fund Grants	www.parks.ca.gov/?page_id=21361
DWR	Water Management	Proposition 1 Integrated Regional Water Management Program	www.water.ca.gov/Work-With-Us/Grants-And-Loans
DWR	Water Management	Sustainable Groundwater Planning Grant Program	www.water.ca.gov/Work-With-Us/Grants-And-Loans
DWR	Water Management	Water Use Efficiency Grants Program CalConserve Revolving Fund	www.water.ca.gov/Work-With-Us/Grants-And-Loans
DWR	Agriculture and Working Lands	Water Use Efficiency Grants Program Agricultural Water Conservation	www.water.ca.gov/Work-With-Us/Grants-And-Loans
DWR	Water Management	Flood Emergency Response Grants Program	www.water.ca.gov/Work-With-Us/Grants-And-Loans
DWR	Water Management	Regional Flood Management Planning	www.water.ca.gov/Work-With-Us/Grants-And-Loans
DWR	Water Management	Small Community Flood Risk Reduction Program	www.water.ca.gov/Work-With-Us/Grants-And-Loans
DWR	Water Management	Water-Energy Grant Program	www.water.ca.gov/Work-With-Us/Grants-And-Loans
DWR	Water Management	Water Desalination Grant Program	www.water.ca.gov/Work-With-Us/Grants-And-Loans
DWR	Water Management	Flood Control Subventions Program	www.water.ca.gov/Work-With-Us/Grants-And-Loans
DWR	Water Management	Flood Corridor Program	www.water.ca.gov/Work-With-Us/Grants-And-Loans

(Continued)

(Continued)

Agency	Funding or financing source	Program	URL
HCD	Housing and Community Development	Housing-Related Parks Program ·	www.hcd.ca.gov/grants-funding/active-no-funding/hrpp.shtml
IBank	Disaster Risk Reduction and Green Infrastructure	California Lending for Energy and Environmental Needs Center	www.ibank.ca.gov/cleen-center/
IBank	Infrastructure	Infrastructure State Revolving Fund	www.ibank.ca.gov/infrastructure-state-revolving-fund-isrf-program/
ILG	Resource Search Engine	Sustainability Funding Resources	www.ca-ilg.org/post/sustainability-funding-resources
OES	Disaster Risk Reduction and Resilience	Emergency Management Performance Grants	www.caloes.ca.gov/cal-oes-divisions/grants-management/criminal-justice-emergency-management-victim-services-grant-programs/emergency-management-performance-grant
OPC	Natural Systems and Green Infrastructure	Proposition 1 Grants	www.opc.ca.gov/category/funding-opportunities/
OPC	Natural Systems and Green Infrastructure	Proposition 84 Grants	www.opc.ca.gov/category/funding-opportunities/
SGC	Housing and Community Development	Affordable Housing and Sustainable Communities Program	www.hcd.ca.gov/grants-funding/active-funding/ahsc.shtml
SGC	Housing and Community Development	Transformative Climate Communities (TCC) Program	http://sgc.ca.gov/programs/tcc/
SWRCB	Water Management	Drinking Water State Revolving Fund	www.waterboards.ca.gov/drinking_water/services/funding/SRF.html

Agency	Category	Program	URL
SWRCB	Water Management	Seawater Intrusion Control Program	www.waterboards.ca.gov/water_issues/programs/grants_loans/swic.shtml
SWRCB	Water Management	Water Recycling Program	www.waterboards.ca.gov/water_issues/programs/grants_loans/water_recycling/
SWRCB	Water Management	Clean Water State Revolving Loan Fund	www.waterboards.ca.gov/water_issues/programs/grants_loans/srf/
WCB	Natural Systems and Green Infrastructure	Natural Heritage Preservation Tax Credit Program	https://wcb.ca.gov/Programs
WCB	Natural Systems and Green Infrastructure	Land Acquisition Program	https://wcb.ca.gov/Programs
WCB	Natural Systems and Green Infrastructure	Habitat Enhancement and Restoration Program	https://wcb.ca.gov/Programs
WCB	Natural Systems and Green Infrastructure	Inlands Wetlands Conservation Program	https://wcb.ca.gov/Programs
WCB	Fire and Forest Management	Forest Conservation Program	https://wcb.ca.gov/Programs
WCB	Natural Systems and Green Infrastructure	California Riparian Habitat Conservation Program	https://wcb.ca.gov/Programs
WCB	Natural Systems and Green Infrastructure	Climate Adaptation and Resiliency Program	https://wcb.ca.gov/Programs

Glossary

Adaptation (climate change)   Adaptation is an adjustment in natural or human systems in response to actual or expected climatic stimuli or their effects, which moderates harm or exploits beneficial opportunities (OPR 2017).

Adaptation Path Dependencies   Adaptation path dependencies are a set of sequential decisions made over time that operate to expand or limit options available to inform or resource future adaptation interventions or investments (Webster 2008).

Adaptation Pathways   Adaptation pathways is a planning approach that sequences the implementation of actions over time, to ensure that systems and assets have the capacity and flexibility to adapt to changing social, environmental and economic conditions (Kingsborough, Jenkins, & Hall 2017).

Adaptive Capacity   Adaptive capacity is the ability of individuals, organizations, institutions and systems to recognize change and to adapt to change by implementing and resourcing adaptation decisions and strategies (Adger et al. 2005).

Adaptive Management   Adaptive management is a process of iteratively planning, implementing, assessing and modifying strategies for managing resources in the face of uncertainty and change. Adaptive management involves adjusting approaches in response to observations of their effect and changes in the system brought on by resulting feedback effects and other variables (Caltrans 2012; IPCC 2014).

Average Annual Loss (AAL)   AAL is the expected loss in a year based on an average of losses observed over a given range of prior years. AAL is often referenced as the mean value of a loss exceedance probability distribution (NAIC 2010).

Climate Change   Climate change refers to a change in the state of the climate that can be identified by changes in the mean and/or the variability of its properties, and that persists for an extended period, typically decades or longer. Climate change may be due to natural internal processes or external forcings, such as those forcings consistent with persistent anthropogenic changes in the composition of the atmosphere or in land use (IPCC 2014).

Climate-Informed Planning Parameter A climate-informed planning parameter is a factor that is employed in the design, planning or investment process, that has been scaled to reflect future climate change (OPR 2017).

Climate Scenarios Climate scenarios are plausible and often simplified representations of the future climate, based on an internally consistent set of climatological relationships that have been constructed for explicit use in investigating the potential consequences of anthropogenic climate change, often serving as input to impact models. Climate projections often serve as the raw material for constructing climate scenarios, but climate scenarios usually require additional information such as the observed current climate (IPCC 2014).

Climate Sensitivity (atmospheric science) Climate sensitivity is a metric used to evaluate the response of the global climate system to a given forcing generally associated with an equilibrium of global mean surface temperature change following a given concentration of atmospheric greenhouse gases (IPCC 2007).

Climate Sensitivity (social and applied science) Climate sensitivity is the degree to which a system, population, object or asset is positively or negatively effected by any given climate change impact or combination of impacts.

Climate Services Climate Services may be defined as scientifically based information and products that enhance users' knowledge and understanding about the impacts of climate on their decisions and actions (AMS 2012).

Co-Benefits Co-benefits are the positive effects that a policy, project or investment aimed at one objective might have on other objectives. Co-benefits may be social, economic or environmental and may also be referred to as ancillary benefits or secondary benefits (LSE 2016).

Community Resilience Community resilience is the ability of communities to withstand, recover and learn from past disasters, and to learn from past disasters to strengthen future response and recovery efforts. This can include but is not limited to physical and psychological health of the population, social and economic equity and well-being of the community, effective risk communication, integration of organizations (governmental and nongovernmental) in planning, response and recovery, and social mobilization for resource exchange, cohesion, response and recovery (IPCC 2014; OPR 2017).

Conduits (financial)   Conduits are governmental or semi-governmental agencies that issue municipal securities or aggregate other sources of funding to finance third-party investments. Conduit borrowers may be not-for-profit organizations, other governmental entities or for-profit private enterprises (MSRB 2018).

Cost-Effectiveness Analysis (CEA)   CEA is a method for choosing among alternatives, based on relative costs and outcomes, in order to select those that are able to most effectively accomplished a pre-determined objective (Levin & McEwan 2000).

Disadvantaged Communities   Disadvantaged communities are those that are disproportionately affected by environmental pollution and other hazards that can lead to negative public health effects, exposure or environmental degradation, or are populations otherwise characterized by low income, high unemployment, low levels of homeownership, high rent burden or low levels of educational attainment, among other sensitivities (California Health & Safety Code § 39711).

Discount Rate   The discount rate is the minimum rate of return required from an investment project to make the investment socially and/or economically desirable to implement (Gollier 2012).

Downscaling (climate modeling)   Downscaling is a method for obtaining high-resolution climate or climate change information from relatively coarse-resolution global climate models (Mearns 2009).

Ecological Resilience   Ecological resilience is the capacity of natural systems subject to instability to absorb disturbances without undergoing change to a fundamentally different stability domain. The general focus is on persistence, change and unpredictability, and the concept is often measured by the magnitude of disturbance that can be absorbed before the system changes its controlling variables (Davidson et al. 2016).

Ecosystem Services   Ecosystem services are the quantitative and qualitative human and environmental benefits that ecosystems provide.

Engineering Resilience   Engineering resilience is the capacity of an engineered system to return to and maintain local stability near an equilibrium state following a perturbation and is measured by the speed of return, as well as the costs associated with identifying, diagnosing, prognosing and resourcing the system's elastic capacity to recover (Davidson et al. 2016).

Environmental Justice   Environmental justice means the fair treatment of people of all races, cultures and incomes with respect to the development, adoption, implementation and enforcement of environmental laws, regulations and policies (California Government Code § 65040.12[e]).

Equity (climate) Climate equity is a discourse that focuses on: (i) addressing the unequal impacts and responses to climate impacts; (ii) identifying who is responsible for causing climate change and for actions to limit its effects; and, (iii) understanding the ways in which climate policies or investment intersects with other dimensions of human development, both globally and domestically (WRI 2014).

Equity (climate economics) Equity in climate economics is defined by the just and open inclusion into a society of all potential participants for the purposes of determining a fair and just distribution of society's resources to mitigate and adapt to climate change (Page 2007).

Equity (financial) Cash, interests and other contributions made by the promoter, grantee or debtor that make up the difference between funds sourced from debt and/or grants and the total contributions necessary to fund a project or program (Keenan 2018).

Expected Value (EV) EV is determined by summing all probable outcomes multiplied by the probability that each outcome will occur.

Exposure (credit) Exposure is the cumulative amount of risk during the life of an investment or financial instrument (Jorion 2007).

Exposure (risk) Exposure refers to the inventory of elements and assets in an area in which hazard events may occur (UNISDR 2017).

Extreme Events (climate) Extreme events are the occurrence of a value of a weather or climate variable above (or below) a threshold value near the upper (or lower) ends of the range of observed values of the variable (IPCC 2014).

First Costs First costs are those incurred in the initial acquisition or development of an asset, including costs associated with the financing of investment but not including maintenance, operations and repair costs (USGSA 2017).

Global Climate Models Global climate models are a numerical representation of the climate system that is based on the physical, chemical and biological properties of its components, their interactions and feedback processes, and that accounts for all or some of its known properties (IPCC 2012).

Integrated Climate Actions Integrated climate actions are program, plans, policies or investments that simultaneously reduce greenhouse gas emissions and decrease the risks posed by climate change on the system where the action is implemented (OPR 2017).

Life-Cycle Cost Accounting (LCCA) Life-Cycle Cost Accounting (LCCA) is an economic method of project evaluation in which all of the present value costs arising from owning, operating, maintaining and ultimately disposing of a project are considered to be potentially important to that decision (NIST 1995).

Maladaptation Maladaptation is defined as actions or inactions that may lead to increased risk of adverse climate-related outcomes; increased vulnerability to climate change; or diminished welfare, now or in the future (IPCC 2014).

Mitigation (climate change) Climate change mitigation is defined as human intervention to reduce the human impact on the climate system; it includes strategies to reduce greenhouse gas sources and emissions and enhancing greenhouse gas sinks (EPA 2013).

Mitigation (disaster risk or hazard) Disaster risk mitigation is the prevention or lessening of the potential adverse impacts of physical hazards through actions that reduce hazard, exposure and vulnerability (IPCC 2014).

Natural and Green Infrastructure Natural and green infrastructure is infrastructure that advances the preservation or restoration of ecological systems through a utilization of engineered systems that use ecological processes to increase resiliency to climate change, manage other environmental hazards or both. This may include, but is not limited to, floodplain and wetlands restoration or preservation; combining levees with restored natural systems to reduce flood risk; and, urban tree planting to mitigate high heat days (California Government Code § 65302).

Net Present Value (NPV) The NPV is a sum of present values of positive and negative cash-flows discounted to the required rate of return for purposes of comparing an initial investment.

Non-market Costs Non-market costs are the costs associated with non-market impacts of a project. These impacts may be quantified and monetized using non-market valuation methods, such as damage cost estimation, prevention cost estimation, hedonic methods, travel cost methods or contingent valuation methods (DWR 2008).

Organizational Resilience Organizational resilience is the capacity of an organization to identify, diagnose, prognose, resource and manage external perturbations in the advancement of the continuity of operations and regular business activities (Sahebjamnia, Torabi, & Mansouri 2015).

Payback Period The payback period is the time required to recoup the investment through an evaluation of discounted or undiscounted cash-flows (NIST 2017).

Portfolio Analysis (PA) PA is a quantitative method for selecting an optimal portfolio that can strike a balance between maximizing the return and minimizing the risk in various uncertain environments (Huang 2010).

Present Value (PV) PV is a factor utilized to determine the value of future streams of income, as discounted by a rate of return often referenced to rates of annualized inflation.

Real Options Analysis (ROA) ROA is a decision-making framework that accounts for uncertainty and provides policymakers managerial and policy flexibility to modify strategies under uncertain conditions. ROA strives to limit the downside of making a wrong decision, allowing policymakers to continue to build or modify strategies as more information becomes available (Buurman & Babovic 2016).

Resilience (climate) Resilience is defined as a capability to anticipate, prepare for, respond to and recover from significant multi-hazard threats with minimum damage to social well-being, the economy and the environment (USGCRP 2018a).

Return on Investment (ROI) ROI is a financial ratio measured by net income over the total costs of an investment for purposes of allowing an investor to calculate the net benefit of an investment.

Risk Risk is a function of the probability of an adverse event times the magnitude of impact of that event. *See also* Uncertainty.

Robust Decision-Making (RDM) RBD is an analytical framework that helps identify potential strategies, characterize vulnerabilities and evaluate trade-offs. RBD is often used to evaluate and develop strategies in areas characterized by uncertainty (RAND 2018).

Scenario Planning Scenario Planning is defined as an ordered process by which actors stress-test – often through complex narratives – what are assumed to be internally consistent perceptions and assumptions about possible futures and the extent to which those futures challenge and shape assumptions and emergent strategies (Lindgren & Bandhold 2003).

Sea Level Rise Sea level rise is the worldwide average rise in mean sea level due to a number of different causes, such as the thermal expansion of sea water and the addition of water to the oceans from the melting of glaciers, ice caps and ice sheets (EPA 2013).

Social Equity *See* equity (climate) or equity (climate economics).

Uncertainty (statistical) Uncertainty is defined as outcomes or impacts of known or unknown events where no probability exists as to the occurrence or impact of such events. *See also* Risk.

Underwriting (financial) Underwriting is the processing of assessing the economic or social viability, feasibility or desirability of any given investment or investment alternative.

Value-at-Risk (VaR) VaR is a collection of various quantitative methodologies for applying probability to determine market and credit risk exposure of an institution. The output of this analysis is an estimate of the maximum loss that can occur with x% confidence over a holding period of t days (Choudhry 2013).

Vulnerability Vulnerability is defined by conditions determined by physical, social, economic and environmental factors or processes which increase the susceptibility of an individual, a community, assets or systems to the impacts of hazards (UNISDR 2017). Vulnerability is often understood to be the outcome of a combination of exposure, sensitivity and adaptive capacity (O'Brien et al. 2007).

Vulnerable Populations   Vulnerable populations include, but are not limited to, women; racial or ethnic groups; low-income individuals and families; individuals who are incarcerated or have been incarcerated; individuals with disabilities; individuals with mental health conditions; children; youth and young adults; seniors; immigrants and refugees; individuals who are limited English proficient (LEP); and Lesbian, Gay, Bisexual, Transgender, Queer, and Questioning (LGBTQQ) communities or combinations of these populations (California Health & Safety Code § 131019.5; OPR 2017).

Index